Maths

Higher

Complete Revision and Practice

Rob Kearsley Bullen, Graham Lawlor

Published by BBC Active, an imprint of Educational Publishers LLP, part of the Pearson Education Group Edinburgh Gate, Harlow, Essex CM20 2JE, England

Text Copyright © Rob Kearsley Bullen and Graham Lawlor 2002, 2007

Design and Concept Copyright © BBC Active 2008, 2010

Designed by specialist publishing services ltd

BBC logo © BBC 1996. BBC and BBC Active are trademarks of the British Broadcasting Corporation.

ISBN 978-1-4066-5459-2

Printed in China (CTPSC/03)

First published 2002

This edition 2010

10 9 8 7 6 5 4 3

Minimum recommended system requirements
PC: Windows(r), XP sp2, Pentium 4 1 GHz processor (2 GHz for Vista), 512 MB of RAM (1 GB for Windows Vista), 1 GB of free hard disk space, CD-ROM drive 16x, 16 bit colour monitor set at 1024 x 768 pixels resolution
MAC: Mac OS X 10.3.9 or higher, G4 processor at 1 GHz or faster, 512 MB RAM, 1 GB free space (or 10% of drive capacity, whichever is higher), Microsoft Internet Explorer® 6.1 SP2 or Macintosh Safari™ 1.3, Adobe Flash® Player 9 or higher, Adobe Reader® 7 or higher, Headphones recommended

If you experiencing difficulty in launching the enclosed CD-ROM, or in accessing content, please review the following notes:
1 Ensure your computer meets the minimum requirements. Faster machines will improve performance.
2 If the CD does not automatically open, Windows users should open 'My Computer', double-click on the CD icon, then the file named 'launcher.exe'. Macintosh users should double-click on the CD icon, then 'launcher.osx'
Please note: the eDesktop Revision Planner is provided as-is and cannot be supported.
For other technical support, visit the following address for articles which may help resolve your issues:
http://centraal.uk.knowledgebox.com/kbase/

If you cannot find information which helps you to resolve your particular issue, please email: Digital.Support@pearson.com.
Please include the following information in your mail:
- Your name and daytime telephone number.
- ISBN of the product (found on the packaging.)
- Details of the problem you are experiencin̄ ̄ ̄ ̄ ̄ ̄ ̄ ̄ ̄ ̄ ̄ tc.
- Details of your computer (operating syster

Contents

Number

Algebra

Shape, space and measures

Handling data

* Only available in the CD-ROM version of the book.

Exam board specification map

Based on the Mathematics National Curriculum for England only
Provides a quick and easy overview of the topics you need to study for the examinations you will be taking.

Topics	Page	AQA A linear	AQA B modular	Edexcel linear	Edexcel modular	OCR A linear	OCR B modular
Number							
The decimal number system	2	✓	✓	✓	✓	✓	✓
Fraction calculations	4	✓	✓	✓	✓	✓	✓
Fractions, decimals and percentages	6	✓	✓	✓	✓	✓	✓
Powers and roots	8	✓	✓	✓	✓	✓	✓
Surds	10	✓	✓	✓	✓	✓	✓
Standard index form	12	✓	✓	✓	✓	✓	✓
Ratio and proportion	14	✓	✓	✓	✓	✓	✓
Percentage calculations	16	✓	✓	✓	✓	✓	✓
Algebra							
Algebraic expressions	18	✓	✓	✓	✓	✓	✓
Formulae and substitution	20	✓	✓	✓	✓	✓	✓
Rearranging formulae	22	✓	✓	✓	✓	✓	✓
Using brackets in algebra	24	✓	✓	✓	✓	✓	✓
Factorising quadratic expressions	26	✓	✓	✓	✓	✓	✓
Algebraic fractions	28	✓	✓	✓	✓	✓	✓
Solving equations	30	✓	✓	✓	✓	✓	✓
Equations of proportionality	32	✓	✓	✓	✓	✓	✓
Trial and improvement	34	✓	✓	✓	✓	✓	✓
Quadratic equations	36	✓	✓	✓	✓	✓	✓
Functions	38	✓	✓	✓	✓	✓	✓
Inequalities and regions	40	✓	✓	✓	✓	✓	✓
Number patterns and sequences	42	✓	✓	✓	✓	✓	✓
Sequences and formulae	44	✓	✓	✓	✓	✓	✓
Lines and equations	46	✓	✓	✓	✓	✓	✓
Curved graphs	48	✓	✓	✓	✓	✓	✓
Transforming graphs	50	✓	✓	✓	✓	✓	✓
Linear simultaneous equations	52	✓	✓	✓	✓	✓	✓
Mixed simultaneous equations	54	✓	✓	✓	✓	✓	✓

Introduction

How to use GCSE Bitesize Complete Revision and Practice

Begin with the CD-ROM. There are five easy steps to using the CD-ROM – and to creating your own personal revision programme. Follow these steps and you'll be fully prepared for the exam without wasting time on areas you already know.

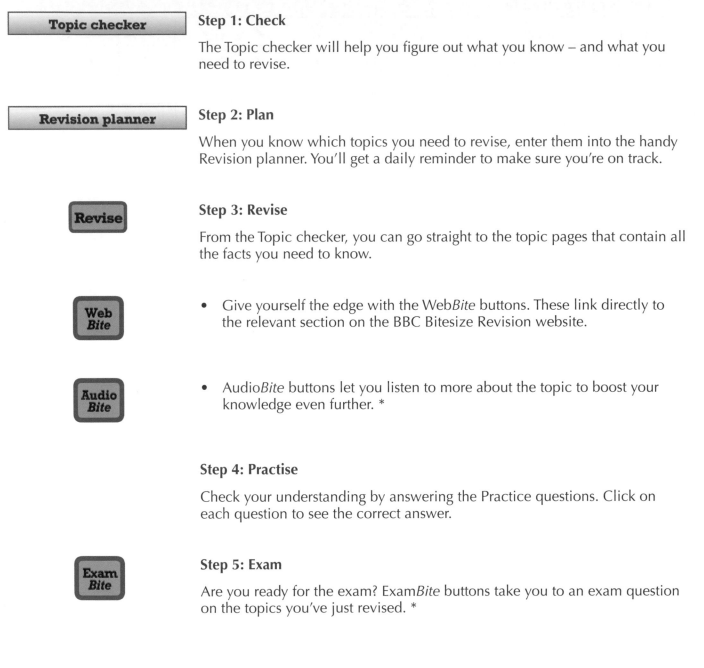

Topic checker

Step 1: Check

The Topic checker will help you figure out what you know – and what you need to revise.

Revision planner

Step 2: Plan

When you know which topics you need to revise, enter them into the handy Revision planner. You'll get a daily reminder to make sure you're on track.

Revise

Step 3: Revise

From the Topic checker, you can go straight to the topic pages that contain all the facts you need to know.

Web Bite

- Give yourself the edge with the Web*Bite* buttons. These link directly to the relevant section on the BBC Bitesize Revision website.

Audio Bite

- Audio*Bite* buttons let you listen to more about the topic to boost your knowledge even further. *

Step 4: Practise

Check your understanding by answering the Practice questions. Click on each question to see the correct answer.

Exam Bite

Step 5: Exam

Are you ready for the exam? Exam*Bite* buttons take you to an exam question on the topics you've just revised. *

* Not all subjects contain these features, depending on their exam requirements.

Interactive book — You can choose to go through every topic from beginning to end by clicking on the Interactive book and selecting topics on the Contents page.

Exam questions — Find all of the Exam questions in one place by clicking on the Exam questions tab.

Last-minute learner — The Last-minute learner gives you the most important facts in a few pages for that final revision session.

 You can access the information on these pages at any time from the link on the Topic checker or by clicking on the Help button. You can also do the Tutorial which provides step-by-step instructions on how to use the CD-ROM and gives you an overview of all the features available. You can find the Tutorial on the Home page when you click on the Home button.

Other features include:

 Click on the draw tool to annotate pages. N.B. Annotations cannot be saved.

 Click on Page turn to stop the pages turning over like a book.

 Click on the Single page icon to see a single page.

 Click on this arrow to go back to the previous screen.

 Click on Contents while in the Interactive book to see a contents list in a pop-up window.

 Click on these arrows to go backward or forward one page at a time.

 Click on this bar to switch the buttons to the opposite side of the screen.

Click on any section of the text on a topic page to zoom in for a closer look.

N.B. You may come across some exercises that you can't do on-screen, such as circling or underlining, in these cases you should use the printed book.

About this book

Use this book whenever you prefer to work away from your computer.
It consists of two main parts:

 A set of double-page spreads, covering the essential topics for revision from each area of the curriculum. Each topic is organised in the following way:

- a summary of the main points and an introduction to the topic

- lettered section boxes cover the important areas within each topic

- key facts highlighting essential information in a section or providing tips on answering exam questions

- practice questions at the end of each topic to check your understanding.

A number of special sections to help you consolidate your revision and get a feel for how exam questions are structured and marked. These extra sections will help you to check your progress and be confident that you know your stuff. They include:

- Topic checker – quick questions covering all topic areas

- a selection of exam-style questions and worked model answers and comments to help you get full marks

- Complete the facts – check that you have the most important ideas at your fingertips

- Last-minute learner – the most important facts in just a few pages.

About your exam

Get organised
You need to know when your exams are before you make your revision plan. Check the dates, times and locations of your exams with your teacher, tutor or school-office.

On the day
Aim to arrive in plenty of time, with everything you need: several pens, pencils, a ruler, and possibly mathematical instruments and a calculator.

On your way or while you're waiting, read through your Last-minute learner.

In the exam room
When you are issued with your exam paper, you must not open it immediately. However, there are some details on the front cover that you can fill in (your name, centre number, etc.) before you start the exam itself. If you're not sure where to write these details, ask one of the invigilators (teachers supervising the exam).

When it's time to begin writing, read each question carefully. Remember to keep an eye on the time.

Finally, don't panic! If you have followed your teacher's advice and the suggestions in this book, you will be well-prepared for any question in your exam.

Techniques to help you remember

There are things that you can do to enhance your memory as you revise.

- **Shapes** – Suppose that you had to remember five different types of numbers; for example, primes, squares, cubes, triangular numbers, and the Fibonacci sequence.

  One easy way to remember these five facts is to draw them around a pentagon:

  You can then add extra diagrams or written facts to the pentagon. The act of drawing the pentagon and labelling it causes you to organise your thoughts in the right way for the topic you're working on; for example, the fact that the pentagon has five sides means that you know you have to remember five facts.

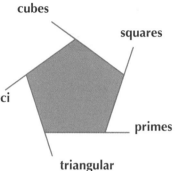

cubes

squares

Fibonacci

primes

triangular

  Suppose you had five categories of information to memorise and each category had three sets of facts. You could develop your shape memory-jogger to look like the one we've drawn on the left.

  Here, the five facts can be listed around the pentagon and each subsection of three facts can be listed around each small triangle. If some of the sub-sections have only one fact, instead of using a triangle, use a circle. The shape memory-jogger is fun, creative and uses both sides of your brain.

- **Chunking** – This means grouping facts together. For instance, one of the easiest ways to remember phone numbers is to group the digits in threes. So, if you can group material, it will make it easier to remember.

- **Acronyms** – Acronyms are 'words' made from the initial letters of facts you need to remember. You can find a number of these (such as BIDMAS) in the book already, but you can easily make your own.

- **Phrases** – Sometimes you can remember a fact using a humorous phrase. For example, people remember that < means 'less than' and > means 'greater than' using the phrase 'the crocodile's mouth eats the biggest thing it can'.

- **Images** – Making images is creative, involving you more actively in the process of revising, helping you to remember things easily. In your revision notes, make pictures and images as often as you can.
  Good images:
  – are colourful
  – are lively and dynamic
  – can make you laugh
  – often have exaggerated aspects
    to them.

  Here's an example illustrating 'along the corridor before you climb the stairs' for co-ordinates:

Topic checker

Go through these questions after you've revised a group of topics, putting a tick if you know the answer.

You can check your answers on pages xv–xix.

Number

1 What is a prime number? ☐

2 Write 120 as a product of its prime factors. ☐

3 What is the Fibonacci Sequence? ☐

4 Evaluate x^0. ☐

5 What is the rule of indices that is used when multiplying two powers of the same base? ☐

6 Evaluate $25^{\frac{1}{2}}$. ☐

7 What is standard index form? ☐

8 Write the following in standard form. **(a)** 1 000 000 **(b)** 0.000 000 000 12 ☐

9 Simplify $\sqrt{12}$. ☐

10 Use your calculator to find $31 \div 13$. Round the answer to:
(a) 1 decimal place **(b)** 3 decimal places **(c)** 3 significant figures. ☐

11 For the numbers 48 and 40, find:
(a) the highest common factor **(b)** the lowest common multiple. ☐

12 Write the ratio 6 : 10 in: **(a)** simplest form **(b)** the form $n : 1$ **(c)** the form $1 : n$. ☐

13 Divide £150 in the ratio 3 : 2 : 1. ☐

14 Calculate the VAT (at 17.5%) on an item costing £30. ☐

15 What is the reciprocal of: **(a)** 5 **(b)** $\frac{1}{8}$ **(c)** $\frac{2}{3}$? ☐

16 Evaluate 2^{-3}. ☐

17 A college had 800 registered students. Last year this increased by 15%. This year it increased by 5%. How many students now attend the college? ☐

18   Calculate: **(a)** $\frac{2}{5} + \frac{3}{8}$ **(b)** $\frac{7}{12} - \frac{3}{10}$ **(c)** $2\frac{1}{2} \cdot \frac{5}{8}$.

19   Rationalise the denominator of $\frac{5}{\sqrt{3}}$.

20   Multiply out and simplify $(1 + 3\sqrt{10})(3 - \sqrt{10})$.

Algebra

21   Solve $6y + 10 = 52$.

22   Solve the inequality $3y - 12 \geqslant 18$.

23   What is the gradient and y-intercept of the graph of $3x + y = 5$?

24   How do you find the equation of a line parallel to a given line, if you know its y-intercept?

25   Solve the simultaneous equations: $2x + y = 13$, $5x + 2y = 29$.

26   What is the highest common factor of $6x^2 + 36x$?

27   What is the lowest common multiple of $5x$ and $6x$?

28   Expand and simplify $(x + 3)(x + 9)$.

29   Expand $(x - 8)^2$.

30   What is a difference of two squares and why is it special?

31   Find the fourth term in the sequence $u_n = 4n + 5$.

32   Find a formula for this sequence: 6, 11, 16, 21, 26, …

33   Rearrange $V = \sqrt{\frac{t}{\pi}}$ to make t the subject.

34   What is a quadratic equation?

35   Factorise $x^2 - 9$.

36   By using trial and improvement, solve $x^3 + x^2 = 8$, correct to one decimal place.

37   Simplify $d^6 \times d^5$.

Topic checker

38   Simplify $2t^6 \div t^3$.   ☐

39   Simplify $(3y^2)^5$.   ☐

40   Solve $h^{\frac{1}{2}} = 25$.   ☐

41   Factorise $x^2 - 8x$.   ☐

42   Factorise $169z^2 - 225y^2$.   ☐

43   Simplify $6x \div \frac{1}{x}$.   ☐

44   Simplify $3(x + 6) \times 4(x + 7)$.   ☐

45   Solve $25h^2 = 100$.   ☐

46   Solve $m^2 + 3m - 12 = -2$.   ☐

47   Solve the simultaneous equations: $x^2 + y^2 = 20$, $2x - y = 0$.   ☐

48   If $f(x) = x^3 + 1$, find: **(a)** $f(4)$ **(b)** $f(9)$.   ☐

49   If $f(x) = 3x^2$, write an expression for $f(5x)$.   ☐

50   Find the two missing numbers from this sequence: 2, 4, 10, ?, 34, ?, …   ☐

51   Find the two missing numbers from this sequence: 10, 5, ?, ?, 0.625, …   ☐

52   Make a the subject of the formula $p = \frac{a}{1-a}$.   ☐

53   Find a formula for the nth term of the sequence: 3, 12, 27, 48, 75, …   ☐

Shape, space and measures

54   By completing the square, solve $x^2 - 8x + 11 = 0$.   ☐

55   Use the quadratic formula to solve $2x^2 + 5x - 6 = 0$.   ☐

56   What are the tests for congruence of triangles?   ☐

57 Are these two shapes congruent? Explain your answer.

58 Find the area of the triangle. $b = 9\text{cm}$, $30°$, $a = 12\text{cm}$

59 Find x. 12, 55°, 15, x

60 What is a translation?

61 What happens when a shape is enlarged by a negative scale factor?

62 Given $\mathbf{m} = \left(\begin{smallmatrix} 8 \\ -3 \end{smallmatrix}\right)$ and $\mathbf{n} = \left(\begin{smallmatrix} -2 \\ -6 \end{smallmatrix}\right)$, find: **(a)** $\mathbf{m} - \mathbf{n}$ **(b)** $3\mathbf{m}$ **(c)** $3\mathbf{m} - 2\mathbf{n}$.

63 What is the acute angle with the same sine as 150°?

64 What is the equation of the line that is parallel to $y = 2x + 1$ and passes through the point (0, 4)?

65 A line joins the points (1, 2) and (3, 7). What is the equation of the line?

66 What is the formula to find the area of a trapezium?

67 What is the size of the angle in a semi-circle?

68 What can you say about the angle subtended by an arc at the centre of a circle, in comparison to the angle subtended by the same arc at any part of the remaining circumference?

69 What can you say about angles in the same segment?

70 How do you draw a perpendicular bisector to a line?

71 What is the difference between a segment of a circle and a sector of a circle?

72 What is the formula for the volume of a sphere?

73 If a measurement is given as 30 ml, correct to the nearest ml, what range of values is possible?

74 If a journey of 75 km takes 90 minutes, what is the average speed?

75 Which quadrilaterals have two lines of symmetry?

Topic checker

Topic checker

---

76   What are the three angle relationships created by parallel lines, and which of these link equal angles? ☐

---

77   What locus is equidistant from two points? ☐

---

78   Draw a net for a cuboid 5 cm by 2 cm by 1 cm. ☐

---

79   What information do you need to give to describe:
(a) a translation   (b) a reflection   (c) a rotation   (d) an enlargement? ☐

---

80   Solve the equation $\sin x = \frac{1}{2}$, for values of x between 0° and 720°. ☐

---

81   Find the volume of a cylinder with base radius 2 cm and height 7 cm, leaving your answer in terms of π. ☐

Handling data

---

82   What is special about a histogram? ☐

---

83   What is frequency density? ☐

---

84   What two statistical values should you use to compare two sets of data? ☐

---

85   In a pie chart, one sector covers 60°. If this represents 12 people, how many people does the pie chart represent altogether? ☐

---

86   On a scatter diagram, when is it not advisable to use the line of best fit to predict values? ☐

---

87   What is the OR rule in probability? ☐

---

88   A bag contains 5 blue, 4 black and 8 white balls. What is the probability of picking a blue or a white ball? ☐

---

89   A playing card is selected from a fair pack of 52 playing cards.
What is the probability that it is not an Ace? ☐

---

90   What is the AND rule in probability? ☐

---

91   A bag contains 10 balls, 6 are red and 4 are white. Two balls are picked from the bag and NOT replaced. What is the probability that 1 red and 1 white ball are selected? Draw a tree diagram to show this information. ☐

---

92   If you know the probability of an outcome occurring, how do you calculate the probability of it not occurring? ☐

---

93   How is relative frequency defined? ☐

Topic checker answers

Number

1 A number with only two factors and the factors are different

2 $2^3 \times 3 \times 5$

3 A sequence of numbers where the next number is found by adding the two previous numbers, e.g. 1, 1, 2, 3, 5, is made up from $0 + 1 = 1$, $1 + 1 = 2$, $1 + 2 = 3$, $2 + 3 = 5$

4 1

5 The indices are added.

6 5

7 A way of writing very large or very small numbers and making them more managable

8 (a) 1×10^6 (b) 1.2×10^{-10}

9 $2\sqrt{3}$

10 (a) 2.4 (b) 2.385 (c) 2.38

11 (a) 8 (b) 240

12 (a) $3 : 5$ (b) $0.6 : 1$ (c) $1 : 1\frac{2}{3}$

13 £75 : £50 : £25

14 £5.25

15 (a) $\frac{1}{5}$ (b) 8 (c) $\frac{3}{2}$

16 $\frac{1}{8}$

17 966

18 (a) $\frac{31}{40}$ (b) $\frac{17}{60}$ (c) 4

19 $\frac{5\sqrt{3}}{3}$

20 $-27 + 8\sqrt{10}$

Topic checker answers

Algebra

21	$y = 7$
22	$y \geqslant 10$
23	gradient $= -3$, intercept $= 5$
24	Since the line is parallel to a known line, it must have the same gradient and, therefore, must fit the $y = mx + c$ general form.
25	$x = 3$, $y = 7$
26	$6x$
27	$30x$
28	$x^2 + 12x + 27$
29	$x^2 - 16x + 64$
30	It is of the general form $x^2 - y^2$ and is a result of the expansion $(x + y)(x - y)$. The special feature is that in the expansion, the middle terms cancel each other out.
31	21
32	$5x + 1$
33	$t = V^2\pi$
34	It is an equation of the form $ax^2 + bx + c = 0$, or one that can be rearranged into this form.
35	$(x + 3)(x - 3)$
36	$x = 1.7$
37	d^{11}
38	$2t^3$
39	$243y^{10}$
40	$h = 625$
41	$x(x - 8)$
42	$(13z - 15y)(13z + 15y)$
43	$6x^2$
44	$12x^2 + 156x + 504$

45	$h = 2$
46	$m = -5$ or 2
47	$x = 2, y = 4$
48	(a) 65 (b) 730
49	$75x^2$
50	20 and 52
51	2.5 and 1.25
52	$a = \frac{p}{1+p}$
53	$u_n = 3n^2$

Shape, space and measures

54	$x = 4 \pm \sqrt{5}$
55	$x = \frac{-5 \pm \sqrt{73}}{4}$
56	SSS, SAS, AAS, RHS
57	Yes, because one can be rotated so it looks identical to the other.
58	27 cm^2
59	12.75 cm
60	It is a shift in the plane, e.g. $\binom{4}{3}$ will translate a point 4 units in the positive x direction and 3 units in the positive y direction.
61	The image is on the opposite side of the centre of enlargement to the object.
62	(a) $\mathbf{m} - \mathbf{n} = \binom{10}{3}$ (b) $3\mathbf{m} = \binom{24}{-9}$
	(c) $3\mathbf{m} - 2\mathbf{n} = \binom{28}{3}$
63	30°
64	$y = 2x + 4$
65	$y = 2.5x - 0.5$
66	$A = \frac{1}{2}(a + b)h$

67 90°

68 The angle at the centre is twice as big as the angle subtended by the same arc at any point on the remaining part of the circumference.

69 They are equal.

70 (i) Open up a pair of compasses to more than half of the line length.
(ii) Draw arcs on both sides of the line segment by placing the pair of compasses at one end of the line segment.
(iii) Now put the pair of compasses at the other end of the line segment and repeat.
(iv) You should now have two pairs of arcs, one each side of the line, that cross. Join the two crossing points with a straight line – this is the perpendicular bisector.

71 A sector is bounded by part of the circumference and two radii; a segment is bounded by part of the circumference and a chord.

72 $V = \frac{4}{3}\pi r^3$

73 29.5 ml < measurement ≤ 30.5 ml

74 50 km/h

75 Rectangle, rhombus

76 Corresponding angles (equal), alternate angles (equal), allied angles (add to 180°)

77 The perpendicular bisector of the points.

78

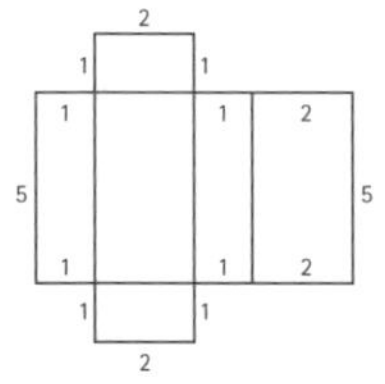

79 **(a)** a vector **(b)** the position of the mirror line **(c)** the angle/direction and the centre of rotation **(d)** the scale factor and the centre of enlargement

80 $x = 30°, 150°, 330°, 510°$

81 28π cm^3

Handling data

82   The frequency is proportional to the area of the bar.

83   It is a measure of the density of the data in class intervals.

84   An average (mean, median or mode) and a measure of spread (range or interquartile range).

85   72

86   When the values are beyond the range of the given data.

87   The OR rule occurs when events have mutually exclusive outcomes. Mathematically, you write this as $P(X \text{ or } Y) = P(X) + P(Y)$.

88 $\frac{13}{17}$

89 $\frac{12}{13}$

90   The AND rule occurs where events have independent outcomes, for instance, rolling a dice and spinning a coin through the air. Mathematically, you write this as $P(X \text{ and } Y) = P(X) \times P(Y)$.

91

$$P(RW) = \tfrac{6}{10} \times \tfrac{4}{9} = \tfrac{4}{15}$$

$$P(WR) = \tfrac{4}{10} \times \tfrac{6}{9} = \tfrac{4}{15}$$

$$P(1 \text{ red, } 1 \text{ white}) = \tfrac{4}{15} + \tfrac{4}{15} = \tfrac{8}{15}$$

92 P (outcome not occuring) $= 1 - P$ (of the outcome occuring)

93   This is the measure of probability based on the past experience of the event, e.g. the probability of an earthquake.

The decimal number system

A Place value

The digit **2** means something different in each of these numbers:

Number	Position of '2'	Value of '2'
125	tens	20
2 130 559	millions	2 000 000
3.28	tenths	$\frac{2}{10}$
0.0026	thousandths	$\frac{2}{1000}$

B Decimal calculations

① **key fact** To multiply a number by 10, 100 or 1000, simply move its digits 1, 2 or 3 places to the left:

$45.67 \times 10 = 456.7$ $45.67 \times 100 = 4567$ $45.67 \times 1000 = 45\,670.$

>> **key fact** To divide by 10, 100 or 1000, move digits 1, 2 or 3 places to the right:

$45.67 \div 10 = 4.567$ $45.67 \div 100 = 0.4567$ $45.67 \div 1000 = 0.045\,67.$

Multiplying by 0.1 is the same as dividing by 10.

② You can use the result of a whole number multiplication to find the answer to many decimal multiplications:

$12 \times 2 = 24,$ so $12 \times 0.2 = 2.4$ $12 \times 0.02 = 0.24$
and $1.2 \times 2 = 2.4$ $1.2 \times 0.2 = 0.24$ $1.2 \times 0.02 = 0.024$, etc.

If one of the numbers is made ten times smaller, the answer will be ten times smaller.

③ A similar rule works with division, but if the number you are dividing by gets smaller, the answer gets bigger:

$12 \div 2 = 6,$ so $12 \div 0.2 = 60$ $12 \div 0.02 = 600$
and $1.2 \div 2 = 0.6$ $1.2 \div 0.2 = 6$ $1.2 \div 0.02 = 60$, etc.

C Negative numbers: addition and subtraction

) **key fact** Adding a negative number gives the same result as subtracting a positive number.

Examples: $4 + (-3) = 4 - 3 = 1$ *brackets are used to make it clearer*
$3 + (-7) = 3 - 7 = -4$ *you could use a number line to help*

Another way to do the last example is to notice that when you swap the numbers in a subtraction, you change the sign of the answer: $7 - 3 = 4$ and so $3 - 7 = -4$.

This can be useful when larger numbers are involved:
$25 + (-42) = 25 - 42$. $42 - 25 = 17$, so $25 - 42 = -17$.

key fact Subtracting a negative number gives the same result as adding a positive number.

Examples: $2 - (-4) = 2 + 4 = 6$.

$(-12) - (-36) = (-12) + 36 = 24$ *use a 'rough' number line to help*

D Negative numbers: multiplication and division

❶ The rules for multiplication and division are very simple:

key fact negative × positive = negative: negative × negative = positive
negative ÷ positive = negative: negative ÷ negative = positive

Examples: $3 \times (-5) = -(3 \times 5) = -15$ $(-4) \times (-10) = 4 \times 10 = 40$
$(-16) \div 2 = -(16 \div 2) = -8$ $(-39) \div (-3) = 39 \div 3 = 13$

❷ You can remember this with the 'word' **SPON**:

'**S**ame (signs) **P**ositive, **O**pposite (signs) **N**egative'. Otherwise, use this table:

×/÷	+	−
+	+	−
−	−	+

❸ **key fact** The square of a negative number is positive.

Example: $(-7)^2 = (-7) \times (-7) = 49$

key fact The cube of a negative number is negative.

Example: $(-2)^3 = (-2) \times (-2) \times (-2) = 4 \times (-2) = -8$

>> practice questions

1 **Do a whole number calculation first, then use the result to answer the question.**

(a) 2.5×3 (b) 0.3×1.2 (c) $6.4 \div 8$ (d) $1.44 \div 0.03$

2 **Work these out without a calculator.**

(a) $3 - 10$ (b) $4 + (-2)$ (c) $13 - (-6)$ (d) $(-5) + 9$

(e) $(-5) + (-9)$ (f) $(-2) - (-7)$ (g) $4 \times (-8)$ (h) $(-3) \times (-7)$

(i) $(-30) \div 6$ (j) $(-24) \div (-3)$ (k) $(-10)^2$ (l) $(-3)^3$

3 **Use your calculator to find these.**

(a) $138 - 272$ (b) $67 + (-125)$ (c) $(-3320) + 2671$ (d) $(-6.93) - (-4.63)$

(e) $(-255) \times 30$ (f) $27 \div (-0.54)$ (g) $(-8) \div (-200)$ (h) $(-0.05)^2$

Fraction calculations

- The numerator is the top number in a fraction and the denominator is the bottom number.

- Create equivalent fractions by multiplying or dividing both numbers by the same thing.

- Change mixed numbers to improper fractions before calculating.

A Equivalent fractions and mixed numbers

1 **key fact** Fractions that contain different numbers, but represent the same amount, are equivalent. Fractions that contain the smallest possible whole numbers are in lowest terms.

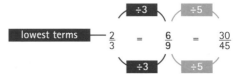

2 **key fact** Fractions in which the numerator is bigger than the denominator are called improper or 'top-heavy'.

Improper fractions can be written as **mixed numbers**: $\frac{7}{3} = \frac{6}{3} + \frac{1}{3} = 2 + \frac{1}{3} = 2\frac{1}{3}$.

Changing mixed numbers to improper fractions is similar: $3\frac{3}{4} = \frac{12}{4} + \frac{3}{4} = \frac{15}{4}$.

3 You can write fractions in order of size by writing them as equivalent fractions with a **common denominator**.

Example: Order these: $\frac{4}{5}, \frac{3}{4}, \frac{7}{12}, \frac{7}{10}, \frac{2}{3}$.

Look at the denominators: 5, 4, 12, 10 and 3 will all divide into 60.

Rewritten, the fractions are $\frac{48}{60}, \frac{45}{60}, \frac{35}{60}, \frac{42}{60}, \frac{40}{60}$.

So the order is $\frac{7}{12}, \frac{2}{3}, \frac{7}{10}, \frac{3}{4}, \frac{4}{5}$.

B Adding and subtracting fractions

1 **key fact** To add or subtract fractions, write them using a common denominator.

If you can find the **lowest** common denominator (LCD), this keeps the numbers small and you are less likely to make a mistake.

2 Sometimes, the LCD will be the denominator of one of the fractions in the question.

Example: $\frac{2}{3} + \frac{4}{9}$ — you can use 9 as the LCD.

$\frac{2}{3} + \frac{4}{9} = \frac{6}{9} + \frac{4}{9} = \frac{10}{9} = 1\frac{1}{9}$.

3 *Example*: $\frac{13}{20} - \frac{5}{12}$

The LCD is 60: $\frac{13}{20} - \frac{5}{12} = \frac{39}{60} - \frac{25}{60} = \frac{14}{60} = \frac{7}{30}$.

C Multiplying and dividing fractions

1 **key fact** To multiply fractions, just multiply the numerators and multiply the denominators. Change mixed numbers to improper form first.

Example: $\frac{3}{4} \times 1\frac{1}{6} = \frac{3}{4} \times \frac{7}{6} = \frac{21}{24} = \frac{7}{8}$.

2 You can keep the numbers small by **cross-cancelling** before you multiply.

Example: $\frac{5}{8} \times \frac{12}{25}$: 5 cancels with 25 and 12 cancels with 8.

3 **key fact** To divide fractions, just invert the second fraction (replace it by its reciprocal) and multiply.

Example: $2\frac{1}{2} \div 1\frac{1}{4} = \frac{5}{2} \div \frac{5}{4} = \frac{5}{2} \times \frac{4}{5} = \frac{2}{1} = 2$.

D Fractions of an amount

1 **key fact** To find a fraction of an amount, divide by the denominator and multiply by the numerator.

Example: What is $\frac{2}{5}$ of £36?

$£36 \div 5 = £7.20 \div 5$ to find $\frac{1}{5}$

$£7.20 \times 2 = £14.40 \times 2$ to find $\frac{2}{5}$

2 **key fact** To express an amount as a fraction of another, write the amounts as a fraction and cancel to lowest terms.

Example: What fraction of 7.5 km is 4.8 km?

$\frac{4.8}{7.5} = \frac{48}{75}$ × top and bottom by 10: now it uses integers

$\frac{48}{75} = \frac{16}{25}$. *cancel*

>> practice questions

1 Write in order of size, smallest to biggest: $\frac{1}{2}, \frac{5}{8}, \frac{7}{12}, \frac{3}{5}, \frac{8}{15}$

2 Calculate the following:

(a) $\frac{7}{10} + \frac{2}{10}$ (b) $\frac{5}{8} - \frac{7}{16}$ (c) $\frac{3}{4} + \frac{4}{5}$ (d) $1\frac{1}{2} - \frac{5}{6}$

(e) $2\frac{3}{8} + 3\frac{1}{2}$ (f) $5 - \frac{2}{9}$ (g) $\frac{1}{3} \times \frac{2}{5}$ (h) $\frac{7}{12} \times \frac{5}{7}$

(i) $\frac{5}{6} \div \frac{2}{3}$ (j) $3\frac{1}{2} \times \frac{5}{7}$ (k) $2\frac{1}{2} \div 1\frac{1}{2}$ (l) $6 \div \frac{3}{4}$

3 Find these amounts:

(a) $\frac{4}{5}$ of 35 km   (b) $\frac{1}{4}$ of 240 ml   (c) $\frac{11}{15}$ of £16.50   (d) $2\frac{1}{2}$ times 450 grams

4 What fraction of the second amount is the first?

(a) 25 cm, 75 cm   (b) £3.60, £5.40   (c) 60 cl, 1 litre   (d) 630 kg, 1.5 tonnes

Fractions, decimals and percentages

- Any quantity can be written as a fraction, a decimal or a percentage.

- Percentages are 'shorthand' for fractions with denominator 100.

- Fractions with denominators 3, 6, 7, 9, 11, 12, ... are equivalent to recurring decimals.

A Changing fractions to decimals

1 If the denominator of a fraction only has the prime factors 2 or 5, its decimal **terminates** (stops after a certain number of digits). Other denominators produce **recurring** decimals.

>> key fact The number of digits in the recurring section is called the **period of the decimal.**

If a group of digits recurs, put a dot over the first and last digit in the recurring part:

$\frac{1}{7} = 0.142\,857\,142\,857\,14\ldots = 0.\overset{\bullet}{1}4285\overset{\bullet}{7}$.

The period is 6.

2 One way to change a fraction to a decimal is to find an equivalent fraction with 10, 100 or 1000 as the denominator. $\frac{2}{5} = \frac{4}{10} = 0.4$

3 **key fact** Remember that a fraction represents a division.

To change $\frac{5}{8}$ to a decimal, divide 5 by 8:

$$\begin{array}{r} 0.\,6\ 2\ 5 \\ 8\overline{)5.\,{}^50^20^40} \end{array}$$

To change $\frac{5}{6}$ to a decimal, divide 5 by 6:

$$\begin{array}{r} 0.\,8\ 3\ 3\ 3\ 3 \\ 6\overline{)5.\,{}^50^20^20^20^20} \end{array}$$

So $\frac{5}{6} = 0.833\,33\ldots$, written $0.8\overset{\bullet}{3}$ to show that the 3 recurs (repeats). The period is 1.

4 It is worth memorising these:

$\frac{1}{3} = 0.\overset{\bullet}{3}, \frac{2}{3} = 0.\overset{\bullet}{6}, \frac{1}{6} = 0.1\overset{\bullet}{6}$.

Note that your calculator will round recurring decimals: $1 \div 6$ might be displayed as $0.166\,666\,67$.

B Changing decimals to fractions

1 The decimal 0.36 means

$\frac{3}{10} + \frac{6}{100} = \frac{30}{100} + \frac{6}{100} = \frac{36}{100}$.

Cancel to lowest terms to get $\frac{9}{25}$.

The quickest way to do this is to look at the number of digits after the decimal point.

If there's one, use 10 as denominator; if two, use 100, etc. Then use the digits as the numerator. So for 0.36, you know that 100 is the denominator and 36 the numerator.

2 If the decimal recurs, use this technique. You need to multiply the decimal by a power of ten based on the period – if it's 1, use 10; if it's 2, use 100, etc. In this section, r stands for the recurring decimal.

- Multiply the decimal by the correct power of ten.

- Subtract the original decimal from this new one. You should get an integer or a terminating decimal.

- Form the fraction. Multiply top and bottom by 10 until the numerator is an integer.

- Cancel to lowest terms.

3 *Example:* Change 0.6̇ to a decimal.

There is 1 recurring digit, so multiply by 10.

$$10r \quad = \quad 6.666666666\ldots$$

$$r \quad = \quad 0.666666666\ldots$$

Subtracting, $9r = 6$

$r = \frac{6}{9} = \frac{2}{3}$ *(lowest terms)*

4 *Example:* Change 0.3̇15̇1̇ to a fraction.

There are 3 recurring digits, so multiply by 1000.

$$1000r \quad = \quad 315.1151151151\ldots$$

$$r \quad = \quad 0.3151151151\ldots$$

Subtracting, $999r = 314.8$

$r = \frac{314.8}{999} = \frac{3148}{9990}$ *multiply by 10 so numerator is an integer*

$r = \frac{1574}{4995}$ *lowest terms*

C Decimals and percentages

1 Decimals and percentages are very closely linked, so it's very easy to change from one to the other.

Be careful with decimals that have only one significant figure:

$0.6 = 60\%$, $0.03 = 3\%$, etc.

>> **key fact** To change a decimal to a percentage, just multiply it by 100.

The first two decimal places become the whole number part of the percentage and any other digits become the decimal part.

So $0.41 = 41\%$, $0.875 = 87.5\%$ (or $87\frac{1}{2}\%$), etc.

2 **key fact** To change a percentage to a decimal, divide by 100.

So $25\% = 0.25$, $6.2\% = 0.062$, $70\% = 0.7$, etc.

$33\frac{1}{3}\% = 33.3\% = 0.3̇$

3 Percentages over 100% become decimals larger than 1.

So $150\% = 1.5$, $200\% = 2$, etc.

>> practice questions

1   **Each column of the table represents one number and contains a fraction, decimal or percentage. Work out what should go in each empty cell. Write fractions in lowest terms.**

	(a)	(b)	(c)	(d)	(e)	(f)	(g)	(h)	(i)	(j)	(k)	(l)
fraction		$\frac{9}{10}$		$\frac{1}{20}$			$\frac{3}{16}$				$\frac{1}{9}$	
decimal			0.35		0.44				2.375	0.072		
percentage	60%					12.5%		0.2%				145%

2   **(a) Change to decimals:**

(i) $\frac{5}{12}$ (ii) $\frac{2}{7}$ (iii) $\frac{1}{9}$

(b) Change to fractions:

(i) 0.4̇   (ii) 0.12̇   (iii) 0.6̇0̇3̇

Powers and roots

Powers are a way of writing repeated multiplication. They consist of a base and an index.

Negative indices produce reciprocals of the numbers produced by positive indices. Fractional indices are equivalent to roots.

To multiply or divide powers of the same base, add or subtract the indices.

A Powers

1 When you multiply repeatedly, you create a **power**.

$3 \times 3 \times 3 \times 3 \times 3 \times 3 = 3^6$.
This is pronounced 'three to the power six'.

>> key fact A power has two parts, a **base** and an **index**.

$$\text{base} \rightarrow a^m \leftarrow \text{index}$$

The base is the number which is being multiplied and the index tells you how many 'copies' of the base to multiply together.

2 Examples:

$$5^2 = 5 \times 5 \quad = 25 \quad \textit{5 squared}$$

$$10^3 = 10 \times 10 \times 10 \quad = 1000 \ \textit{10 cubed}$$

$$2^5 = 2 \times 2 \times 2 \times 2 \times 2 = 32 \quad \textit{2 to the fifth}$$

Notice that 3^1 is just 3.

3 For any base a, $a^0 = 1$. So $2^0 = 1$, $7^0 = 1$, $23.6^0 = 1$, etc.

4 **key fact** Negative powers are the reciprocals of positive powers.

So $3^{-2} = \frac{1}{3^2} = \frac{1}{9}$, $10^{-6} = \frac{1}{10^6} = \frac{1}{1\,000\,000}$,

$7^{-3} = \frac{1}{7^3} = \frac{1}{343}$, etc.

5 Your calculator has keys for finding powers.

It has a key which will calculate any power

 or x^y,

a squaring key $\square^2$

and may even have a cubing key $\square^3$.

Make sure you know how to use them.

B Roots

1 **key fact** Roots are the opposites or inverses of powers.

The diagrams above show that:

$\sqrt{25} = 5$, because $5^2 = 25$

$\sqrt[3]{1000} = 10$, because $10^3 = 1000$

$\sqrt[5]{32} = 2$, because $2^5 = 32$.

2 Your calculator has keys for finding roots.

It has a key which will calculate any root

$\sqrt[\square]{\square}$ or $\sqrt[x]{y}$,

a square root key $\sqrt{\square}$

and may even have a cube root key $\sqrt[3]{\square}$.

Fractional powers are equivalent to roots.

So $5^{\frac{1}{2}} = \sqrt{5}$; $10^{\frac{1}{3}} = \sqrt[3]{1000}$; $32^{\frac{1}{5}} = \sqrt[5]{32}$

When the index is not a unit fraction, the numerator operates as a power.

So $8^{\frac{2}{3}} = \sqrt[3]{8^2} = \sqrt[3]{64} = 4$.

Alternatively, $8^{\frac{2}{3}} = (\sqrt[3]{8})^2 = 2^2 = 4$.

Fractional indices can be negative too!

So $25^{-\frac{5}{2}} = \frac{1}{25^{\frac{5}{2}}} = \frac{1}{(\sqrt{25})^5} = \frac{1}{5^5} = \frac{1}{3125}$.

C Combining powers

key fact **When two powers of the same base are multiplied together, add their indices.**

Example:

$5^2 \times 5^3 = (5 \times 5) \times (5 \times 5 \times 5) =$
$5 \times 5 \times 5 \times 5 \times 5 = 5^5$
so $5^2 \times 5^3 = 5^{(2+3)}$

Check:

$25 \times 125 = 3125$ ✓

key fact **When two powers of the same base are divided, subtract their indices.**

Example:

$10^6 \div 10^4 = 10^{(6-4)} = 10^2$

Check:

$1\,000\,000 \div 10\,000 = 100$ ✓

D The 'power of a power' rule

Sometimes a power can be used as the base for a new power.

$(3^2)^5 = 3^2 \times 3^2 \times 3^2 \times 3^2 \times 3^2 =$
$3^{(2+2+2+2+2)} = 3^{(2\times5)} = 3^{10}$.

This is known as the 'power of a power' rule. This is the only time that you can multiply indices.

You can use it to solve a problem like this:

Find x, if $9^2 = 3^x$

$9 = 3^2$, so $9^2 = (3^2)^2 = 3^{(2\times2)} = 3^4$

So $x = 4$

>> practice questions

1 **Calculate these powers and roots. Round answers to 3 significant figures if required.**

(a) 2^9 (b) 22^2 (c) 5^6 (d) 2^{-5}

(e) $\sqrt{64}$ (f) $\sqrt{7}$ (g) $\sqrt[5]{243}$ (h) $\sqrt[3]{10}$

2 **Use the index rules to find the missing numbers.**

(a) $4^3 \times 4^2 = 4^?$ (b) $10^6 \times 10^4 = 10^?$

(c) $5^7 \div 5^2 = 5^?$ (d) $2^4 \div 2^7 = 2^?$

(e) $(3^2)^4 = 3^?$ (f) $(1.5^3)^3 = 1.5^?$

(g) $\frac{4^4 \times 4^5}{4^7} = 4^?$ (h) $\frac{(10^4)^3}{10^2} = 10^?$

3 **Find the value of x in each case.**

(a) $2^6 = 4^x$ (b) $3^{10} = 9^x$

(c) $2^6 = 8^x$ (d) $25^5 = 5^x$

4 **Write as a number or fraction:**

(a) $16^{\frac{1}{2}}$

(b) $27^{\frac{2}{3}}$

(c) $16^{-\frac{1}{4}}$

(d) $49^{-\frac{3}{2}}$

Surds

> When a number is written as the product of two equal factors, that factor is called the square root of the number, e.g. $4 = 2 \times 2$ means that 2 is the square root of 4 or $\sqrt{4} = 2$.

> Notice, -2 is also a square root of 4, because $-2 \times -2 = 4$.

A Rational numbers and surds

1. Integers and fractions are called **rational** numbers.

   The square roots of some numbers are rational,

   e.g. $\sqrt{4} = 2$, $\sqrt{36} = 6$, $\sqrt{9} = 3$

>> **key fact** $\sqrt{}$ **stands for the positive square root only.**

2. This is **not** true of all square roots. Numbers like $\sqrt{2}$, $\sqrt{7}$ etc. can't be written as fractions, and can't be expressed exactly as a decimal, though you can approximate to as many decimal places as you like. Numbers like this are called **irrational** numbers.

3. The only way to give an exact answer containing these numbers is to leave them in square root form,

   e.g. $\sqrt{2}$. In this form, they are called **surds**.

B Simplifying surds

1. $\sqrt{12}$ can be written as $\sqrt{(4 \times 3)}$, which is the same as $\sqrt{4} \times \sqrt{3}$, or $2\sqrt{3}$.

   The 'trick' is to split the number inside the root into a product, one of whose parts is a perfect square.

   In the same way, $\sqrt{18} = \sqrt{(9 \times 2)} = \sqrt{9} \times \sqrt{2} = 3\sqrt{2}$.

2. **Fractional surds** can sometimes be simplified using a similar technique:
   $$\sqrt{\frac{3}{25}} = \frac{\sqrt{3}}{\sqrt{25}} = \frac{\sqrt{3}}{5}$$
   In this example, an equivalent fraction is used to create a perfect square in the denominator:
   $$\sqrt{\frac{1}{18}} = \sqrt{\frac{2}{36}} = \frac{\sqrt{2}}{\sqrt{36}} = \frac{\sqrt{2}}{6}$$

C Multiplying surds

1. To multiply expressions containing surds, follow the same rules as you would use to expand brackets in algebra.

   Simplify $\sqrt{3}\,(2 + \sqrt{7})$.
   $$\sqrt{3}\,(2 + \sqrt{7}) = \sqrt{3} \times 2 + \sqrt{3} \times \sqrt{7}$$
   $$= 2\sqrt{3} + \sqrt{21}.$$

In the example above, there are no like terms to collect, but when the same surd occurs in each bracket, the expansion can be simplified.

Simplify $(3 + \sqrt{2})(5 + \sqrt{2})$.

$$(3 + \sqrt{2})(5 + \sqrt{2}) = 3(5 + \sqrt{2}) + \sqrt{2}\,(5 + \sqrt{2})$$
$$= 3 \times 5 + 3\sqrt{2} + 5\sqrt{2} + \sqrt{2} \times \sqrt{2}$$
$$= 15 + 8\sqrt{2} + 2$$
$$= 17 + 8\sqrt{2}$$

D Rationalising a denominator

Sometimes, when you calculate an answer, the surd appears on the bottom of a fraction. It is usual to rewrite the answer so that the surd appears on the top instead, with a whole number in the denominator. This is called **rationalising** the denominator.

Rationalise $\dfrac{3}{\sqrt{10}}$

$$\frac{3}{\sqrt{10}} = \frac{3}{\sqrt{10}} \times \frac{\sqrt{10}}{\sqrt{10}} = \frac{3 \times \sqrt{10}}{\sqrt{10} \times \sqrt{10}} = \frac{3\sqrt{10}}{10}$$

This technique involves multiplying top and bottom by the surd found on the bottom.

If the denominator contains a combination of integers and surds such as $3 + 2\sqrt{7}$, create a new term by reversing the sign of the surd (in this case, $3 - 2\sqrt{7}$), then multiply top and bottom by it.

Rationalise $\dfrac{2}{3 + 2\sqrt{7}}$

$$\frac{2}{3 + 2\sqrt{7}} = \frac{2(3 - 2\sqrt{7})}{(3 + 2\sqrt{7})(3 - 2\sqrt{7})} = \frac{6 - 4\sqrt{7}}{9 - 6\sqrt{7} + 6\sqrt{7} - 4 \times 7} = \frac{6 - 4\sqrt{7}}{9 - 28} = \frac{6 - 4\sqrt{7}}{-19} = \frac{4\sqrt{7} - 6}{19}$$

>> practice questions

Simplify these surd expressions:

1 $\sqrt{32}$ 2 $\sqrt{300}$ 3 $\sqrt{245}$ 4 $\sqrt{\dfrac{3}{16}}$ 5 $\sqrt{\dfrac{8}{9}}$ 6 $\sqrt{\dfrac{1}{50}}$

7 $\sqrt{7}\,(3 + \sqrt{3})$ 8 $\sqrt{2}(1 - \sqrt{2})$ 9 $\sqrt{10}(4 + 3\sqrt{10})$

10 $(2 + \sqrt{6})(1 + \sqrt{6})$ 11 $(4 + \sqrt{3})(3 - 2\sqrt{3})$ 12 $(10 - 2\sqrt{5})(1 - 5\sqrt{5})$

13 $\dfrac{3}{\sqrt{11}}$ 14 $\dfrac{2}{5\sqrt{6}}$ 15 $\sqrt{2} \div \sqrt{6}$

16 $\dfrac{\sqrt{3}}{2 - \sqrt{3}}$ 17 $\dfrac{1 + \sqrt{5}}{1 - \sqrt{5}}$

Standard index form

 Any number can be written in standard index form, but it is particularly useful for very large or very small numbers.

A number is in standard index form when it is written as $a \times 10^n$, where n is an integer and a is between 1 and 10.

Large numbers have positive indices, small numbers (less than 1) have negative indices.

A Powers of ten

1 Here are some of the powers of ten:

10^{-6}	10^{-3}	10^{-2}	10^{-1}	10^0	10^1	10^2	10^3	10^6
0.000001	0.001	0.01	0.1	1	10	100	1000	1 000 000

These are used to write numbers in standard index form (also called *standard form* or *scientific notation*).

2 **key fact** Numbers in standard index form consist of a number between 1 and 10, multiplied by a power of ten.

3 Large numbers:

200 000	$= 2 \times 100\,000$	$= 2 \times 10^5$
4 250 000	$= 4.25 \times 1\,000\,000$	$= 4.25 \times 10^6$
560	$= 5.6 \times 100$	$= 5.6 \times 10^2$

4 Small numbers:

0.002	$= 2 \times 0.001$	$= 2 \times 10^{-3}$
0.000 073 3	$= 7.33 \times 0.000\,01$	$= 7.33 \times 10^{-5}$
0.89	$= 8.9 \times 0.1$	$= 8.9 \times 10^{-1}$

B Multiplication and division

1 To multiply two numbers in standard index form, rearrange them so the number and power parts can be dealt with separately (the brackets used here are just to make it easier to read).

$$(3 \times 10^4) \times (4 \times 10^7) = 3 \times 4 \times 10^4 \times 10^7$$
$$= 12 \times 10^{11}$$

This number is in index form, but not in *standard* index form, because the first part is bigger than 10. You need to adjust the size of the first number, then compensate with the power to keep the value the same:

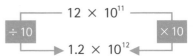

2 To divide two numbers in standard index form, follow similar steps.

$$(1.2 \times 10^8) \div (2 \times 10^3) = 1.2 \div 2 \times 10^8 \div 10^3$$
$$= 0.6 \times 10^5$$

This also needs to be converted to standard index form.

C Addition and subtraction

You can add or subtract numbers in standard index form, as long as the indices in the numbers are the same (it's a bit like using common denominators when you add fractions – the denominators have to match).

Examples: $(2 \times 10^5) + (4 \times 10^5) = 6 \times 10^5$

$(2.9 \times 10^9) - (2.4 \times 10^9) = 0.5 \times 10^9 = 5 \times 10^8$

the answer had to be converted to standard index form

If the indices are different, you need to adjust one of the numbers to match the other. It's usually best to keep the largest index as it is.

Examples:

So $(8 \times 10^3) + (5 \times 10^4) = (0.8 \times 10^4) + (5 \times 10^4) = 5.8 \times 10^4$.   *adjust the first number*

$(3 \times 10^{-6}) - (1.4 \times 10^{-7}) = (3 \times 10^{-6}) - (0.14 \times 10^{-6}) = 2.86 \times 10^{-6}$ *adjust the second number*

D Using your calculator

Your calculator should have a key for entering numbers in standard index form.

It should look like one of these:

To enter 5×10^9, key in 5 EXP 9 .

To enter 2×10^{-4}, key in

Once a number is entered, use it as you would a normal number. Try out some of the calculations from sections B and C to check that you're entering numbers correctly.

Never, under any circumstances, enter

 5   ×   1   0   EXP   9 .

This will give you 50×10^9, which is ten times too large!

>> practice questions

1   Write these numbers in standard index form.

(a) 20 000   (b) 4 000 000

(c) 550 000   (d) 97 100 000 000

(e) 0.05   (f) 0.000 000 3

(g) 0.000 72   (h) 0.355

2   Write these numbers in ordinary form (without indices).

(a) 6×10^7 (b) 3.23×10^5

(c) 1.9×10^{10} (d) 4×10^1

(e) 7×10^{-5} (f) 1.99×10^{-8}

(g) 9.03×10^{-2} (h) 8×10^{-10}

3   Work these out **without** using your calculator. Write your answers in standard index form.

(a) $(2 \times 10^5) \times (4 \times 10^2)$

(b) $(1.6 \times 10^7) \div (3.2 \times 10^5)$

(c) $(4 \times 10^4) + (9 \times 10^3)$

(d) $(5 \times 10^8) - (1.1 \times 10^7)$

4   Use your calculator to answer these. Write your answers in standard index form.

(a) $(1.5 \times 10^5)^2$

(b) $(3.3 \times 10^7) \times (2.1 \times 10^3) + (8 \times 10^8)$

(c) $\dfrac{(7.9 \times 10^5) - (3.3 \times 10^6)}{(2.5 \times 10^3)}$

(d) $\sqrt{(6.25 \times 10^{-8})}$

Ratio and proportion

- Ratios are used to compare amounts using simple numbers.

- Proportional amounts are always in the same ratio.

- Maps and plans use ratios to describe their scales.

A Ratios

1 These ratios compare the numbers of fish in the tank.

There are 12 red fish and 8 green
fish in the tank.
red : green = 12 : 8

For every 3 red fish, there are
2 green fish.
red : green = 3 : 2

2 **key fact** Equivalent ratios,
like equivalent fractions, contain
different numbers but describe
the same relationship.

You create equivalent ratios by multiplying
or dividing all the numbers in a ratio by
the same thing.

3 The equivalent ratio that contains the smallest
whole numbers is in **lowest terms** or **simplest
form**. 3 : 2 is the simplest form of 12 : 8.

4 A ratio that contains a 1 is called a **unitary**
ratio. There are two unitary forms for 3 : 2.

$3 : 2 = 1.5 : 1$ *divide both parts by 2*

$3 : 2 = 1 : \frac{2}{3}$ *divide both parts by 3*

5 Ratios may have more than two parts. For
example, in a recipe you might need 200 g
flour, 100 g sugar and 50 g butter. The ratio is

flour : sugar : butter = 200 : 100 : 50 = 4 : 2 : 1.

B Proportional division

1 You can divide up an amount according to a
ratio. To do this:

- Add up the numbers in the ratio. This tells
  you how many equal parts to divide the
  amount into.

- Divide to find the amount in one part.

- Multiply the amount in one part by each
  number in the ratio. These numbers solve
  the problem.

2 *Example:*

Suppose £75 is to be divided between Sophie
and her mother in the ratio 1 : 4. Follow the
instructions: 1 + 4 = 5 parts.

5 parts = £75
1 part   = £75 ÷ 5 = £15
4 parts = £15 × 4 = £60

So Sophie receives £15 and her mother £60.

C Proportional quantities

1 **key fact** Two quantities are in direct proportion if they stay in a fixed ratio. For example, in these patterns, the ratio of circles to squares is always 2 : 1.

2 When solving problems involving proportional quantities, the **unitary method** is very useful. This involves changing the amount being varied to 1.

Example:

Suppose you can buy 5 litres of petrol for £4.00. How much would 12 litres cost?

5 litres	cost	£4.00	*write down what you know*
1 litre	costs	£4.00 ÷ 5 = £0.80	*change number of litres to 1*
12 litres	cost	£0.80 × 12 = **£9.60**	*multiply to obtain answer*

How much petrol could you buy for £15.00?

£4	will buy	5 litres	*write down what you know*
£1	will buy	5 ÷ 4 = 1.25 litres	*change number of £ to 1*
£15	will buy	1.25 × 15 = **18.75 litres**	*multiply to obtain answer*

3 Currency conversions involve proportional quantities. For example, 1 euro is worth about 90 pence, so euros : pounds = 1 : 0.9 = 1.11 : 1.

D Maps, scale models and plans

1 A popular scale for maps is 2 cm representing 1 km in reality.

The ratio of map sizes to actual sizes is
2 cm : 1 km = 2 cm : 1000 m =
2 cm : 100 000 cm.

This can now be written without units as
2 : 100 000 = 1 : 50 000.

It means that lengths on the map are $\frac{1}{50\,000}$ of

the real lengths, or that real objects are 50 000 times larger than they appear on the map.

2 A scale model is also a fraction or multiple of the size of the real object. For example, if a model plane is made to a scale of 1 : 72 and is 15 cm long, the real plane is 72 × 15 = 1080 cm long, or 10.8 m.

3 The plan of a small electronic component might be 18 cm long, drawn at a scale of 40 : 1. In reality, its size is $\frac{1}{40}$ that of the drawing = 18 ÷ 40 = 0.45 cm.

>> practice questions

1 **Write these ratios in their simplest form.**

 (a) 10 : 2   (b) 9 : 6

 (c) 75 : 100   (d) 30 : 36

2 **Write these ratios in their unitary forms,** $1 : n$ **and** $n : 1$.

 (a) 2 : 5   (b) 10 : 6

3 **Divide £150 in the ratios given.**

 (a) 2 : 3   (b) 9 : 1

4 **100 ml of milk contain 50 calories; how many calories do 250 ml contain?**

5 **A ship that is 120 m long is modelled on a scale of 1 : 250. How long is the model?**

Percentage calculations

- Percentages are fractions with denominator 100.
- Percentage changes involve multiplying or dividing by an appropriate decimal.
- Repeated changes involve the use of powers.

A Percentages of an amount

1 key fact Remember that percentages are just fractions 'out of 100'.

You can use the **unitary method** to calculate a percentage, by finding 1% first.

Example: What is 6% of £350?

$$£350 \div 100 = £3.50 \qquad 6\% = £3.50 \times 6 = £21$$

2 Calculations like that in part 1 can be done 'in one go' on your calculator.

To find $17\frac{1}{2}\%$ of £26, you would enter `1` `7` `.` `5` `÷` `1` `0` `0` `x` `2` `6` `=`.

Alternatively, you could change $17\frac{1}{2}\%$ to a decimal, 0.175, and just multiply £26 by 0.175.

3 If the percentage is the same as a simple fraction, use it to make the calculation easier.

Example: Find 25% of £60. 25% is the same as $\frac{1}{4}$, so 25% of £60 = £60 ÷ 4 = £15.

B Percentage changes

1 key fact Percentage increases occur when things grow or sums of money have interest added.

There are two ways to calculate a percentage increase.

Example: Suppose you could buy 10 litres of petrol for £8, but this is increased by 5%. What is the new price?

Method 1 – Calculate the increase, then add this onto the original price.

5% of £8 = £8 ÷ 100 × 5 = £0.40. New price = £8 + £0.40 = £8.40.

Method 2 – Find the new total percentage, then calculate this.

100% + 5% = 105%. 105% of £8 = £8 × 1.05 = £8.40.

2 key fact Percentage decreases occur when things shrink, or sums of money have tax deducted or are discounted.

Example: Suppose a pair of jeans costing £26 is reduced by 15% in a sale. What is the discounted price?

Method 1 – Calculate the decrease, then subtract this from the original price.

15% of £26 = £26 ÷ 100 × 15 = £3.90. Sale price = £26 − £3.90 = £22.10.

Method 2 – Work out what percentage of the original price is being charged.

100% − 15% = 85%. 85% of £26 = £26 × 0.85 = £22.10.

C Reversing a change

1 To find the amount before a change, you must use *Method 2* demonstrated above.

Example (decrease): Maddy takes home £180 a week after deductions of 25%. What is her gross pay?

She takes home 75% of what she earns, so take-home pay = 0.75 × gross pay.

So gross pay = take-home pay ÷ 0.75 = £180 ÷ 0.75 = £240.

2 *Example (increase)*: In one year, the population of a town increased by 5% to 4725 people. What was the population at the start of the year?

New population = 1.05 × old population.

So old population = new population ÷ 1.05 = 4725 ÷ 1.05 = 4500.

D Repeated changes

When changes are repeated, you can use powers to reduce the working.

Example: £2000 is invested at 4% compound interest. What is it worth after 5 years?

Each year, the amount in the account is multiplied by 1.04, so the amount after 5 years is £2000 × 1.04^5 = £2433.31, to the nearest penny.

1 Calculate the increases and decreases.

	£200	£11	25 litres	500 tonnes
Increase by 10% …				
Increase by 3% …				
Decrease by 20% …				
Decrease by 33% …				

2 Convert the following to percentages.

(a) 28 out of 40 in a test. (b) 4500 people out of 30 000.

(c) 4 cm out of 1 m. (d) £16 out of £50.

3 Calculate the percentage increase or decrease.

(a) £300 changed to £360. (b) £2.50 changed to £2.65.

(c) 500 people changed to 370. (d) 4 000 000 tonnes changed to 3 940 000 tonnes.

4 Calculate the original amounts. Round to the nearest penny when required.

(a) £38.50 after an increase of 10%. (b) £42.30 without VAT at 17.5%.

(c) £65 after a discount of 20%. (d) £2310 after 30% commission deducted.

Algebraic expressions

 Algebraic terms are built from letters and numbers. Expressions are made by combining terms.

 An algebraic statement is two expressions connected by an equals sign. The different types of algebraic statement are formulae, equations and identities.

A Notation

In algebra, letters are used to stand for numbers. A letter may stand for an unknown number, or it may indicate a place where you can use any number you wish.

>> **key fact** **Letters represent variables or unknowns and are usually printed in an *italic* font.**

The standard operations that you can use on numbers apply to letters in algebra too. However, there are some special ways of writing them down.

Multiplication:
$2x$ means '2 times the number x'.
ab means 'multiply the numbers a and b together'.

Division: $\frac{z}{10}$ means 'divide the number z by 10'.

B Terms

A **term** in algebra can be made up from:

- single letters or numbers

- letters and numbers that are linked by multiplication or division.

Examples:

$$x \quad a \quad 1 \quad 25 \quad 2x \quad ab \quad 5x^2 \quad \frac{N}{2} \quad \frac{24xyz}{5ac}$$

C Expressions

An **expression** in algebra can be made up from:

- a single term

- a number of terms linked by adding or subtracting.

Examples:

$$x + c \qquad n - 5 \qquad 2xy + 2zy + 2zx$$
$$\frac{p^2}{2} + \frac{q^2}{3} - \frac{t^2}{10}$$

You can build up more complicated expressions by multiplying or dividing simpler expressions.

Examples:

$$(x + 1)(2x - 3) \qquad \frac{2d + 5}{d + 5}$$

D Formulae

> **key fact** A formula is a set of instructions for calculating something.

The thing being calculated is called the **subject** of the formula and is usually a single letter on the left of the equals sign. The instructions appear as an expression on the right of the equals sign.

 Examples:

- To calculate the perimeter of a rectangle (P) if you know its length (l) and width (w), double the length, double the width and add these together.
  The formula is $P = 2l + 2w$.

- To calculate the average speed of a journey (s) if you know the distance (d) and time taken (t), divide the distance by the time.
  The formula is $s = \dfrac{d}{t}$.

E Equations

> **key fact** An equation is like a mathematical puzzle which requires a solution.

A letter (often x) stands for the unknown number in the equation. You solve the equation by simplifying it according to the rules of algebra.

Examples: $2x + 3 = 11$ (solution: $x = 4$)

$\dfrac{x}{5} = 20$ (solution: $x = 100$)

Formulae can sometimes produce equations to solve. Suppose a rectangle's perimeter is $32\,cm$ and its width is $5\,cm$. Use the formula for the perimeter from section D.

$P = 2l + 2w$. Substitute the known values.

$32 = 2l + 10$. This is now an equation to find l.

F Identities

> **key fact** An identity is a mathematical statement that is true whatever values the letters take. Identities often show you two different ways of writing the same thing.

 Example: You can calculate the perimeter of a rectangle like this: add the width to the length, then double the result.
This would be written $P = 2(l + w)$.

So $2(l + w) = 2l + 2w$. This is an identity, because it is true whatever values you choose for l and w.

>> practice questions

1 How many terms are there in each expression?

   (a) $3x + 5$ (b) ab (c) $P + Q - R$ (d) 1357 (e) $\dfrac{u}{v}$ (f) c^2

   (g) $x^2 + 2xy + y^2$ (h) πr^2 (i) $abcdef$ (j) $t^3 - 3t^2 + 3t - 1$

2 Decide whether each of these algebraic statements is a formula, an equation or an identity.

   (a) $3p + 5 = 20$ (b) $2x + 2x = 4x$ (c) $12 = 10 - x$ (d) $C = \pi d$

   (e) $x^2 + 5x = 0$ (f) $F = \dfrac{9C}{5} + 32$ (g) $(x + y)(x - y) = x^2 - y^2$

   (h) $z + 6 = 2z + 8$ (i) $c^2 = a^2 + b^2$ (j) $(pq)^2 = p^2 q^2$

Formulae and substitution

> The subject of a formula is the thing the formula is designed to calculate.

> Substituting numbers into a formula replaces letters by their values.

A The subject of a formula

① key fact A formula is a set of instructions for calculating something.

The thing being calculated is called the **subject** of the formula and is usually a single letter on the left of the equals sign. The instructions appear as an expression on the right of the equals sign.

② *Example:*

- To calculate the area of a triangle (A) if you know the length of its base (b) and its perpendicular height (h), multiply them together, then halve the result.

The formula is $A = \frac{bh}{2}$.

B Substitution

① To use a formula to carry out the calculation it describes, you need to **substitute** numbers for letters in the formula.

>> key fact Substitution means replacing each known letter in a formula by its value.

② Use the word VASE to help you remember how to set things out:

V	alues	write down what each letter stands for
A	lgebra	write out the formula
S	ubstitute	replace each letter by its value
E	valuate	calculate the result

Example:

The perimeter of a triangle, P cm, whose sides are a cm, b cm and c cm long, is

$$P = a + b + c.$$

Find the perimeter if $a = 4$, $b = 10$ and $c = 7$.

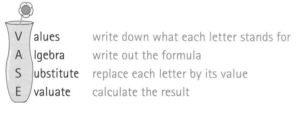

V	$a = 4$, $b = 10$, $c = 7$
A	$P = a + b + c$
S	$= 4 + 10 + 7$
E	$= 21$

③ Be careful when the formula contains multiplication.

For example, if $y = 2x$ and $x = 3$, when you substitute you must remember to put back the 'invisible' multiplication sign: $y = 2 \times 3$, not $y = 23$!

④ Take extra care if any of the numbers you need to substitute are negative. Make sure you obey the rules for working with negative numbers.

Example:

Find the value of R using the formula $R = 5x - 3y$, if $x = 2$ and $y = -4$.

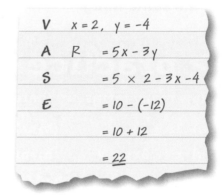

V	$x = 2$, $y = -4$
A	$R = 5x - 3y$
S	$= 5 \times 2 - 3 \times -4$
E	$= 10 - (-12)$
	$= 10 + 12$
	$= 22$

C Creating formulae

1 A formula is a bit like a mathematical recipe.
You need ingredients (the letters and numbers used in the formula) and a method (a way to combine the ingredients).

2 *Example*:

To calculate a Fahrenheit temperature from a Celsius temperature:

'Ingredients': Celsius temperature (C), Fahrenheit temperature (F).

'Method': Multiply the Celsius temperature by 9, then divide by 5. Add 32 to the result.

The formula is $F = \dfrac{9C}{5} + 32$.

3 Always check any formula you create by substituting some suitable numbers into it. In the formula above, you could use the fact that the boiling point of water is 212°F or 100°C.

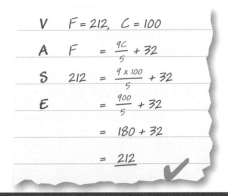

V $F = 212, \ C = 100$

A $F = \dfrac{9C}{5} + 32$

S $212 = \dfrac{9 \times 100}{5} + 32$

E $= \dfrac{900}{5} + 32$

$= 180 + 32$

$= 212$ ✔

practice questions

1 Given that $a = 2$, $b = 3$, $c = 10$, $x = 0.1$ and $y = -4$, evaluate the following expressions.

(a) $a + b + c$ (b) $5c - 8$

(c) $ab + y$ (d) $\dfrac{x}{c}$

(e) $\dfrac{12a - 7b}{cx}$ (f) c^2

(g) $4ay$ (h) $(cy)^2$

(i) $5(ac + by)$ (j) $\sqrt{bc + 3a}$

2 Use the formula $Z = \dfrac{u - v}{u + w}$ to calculate Z for each of these sets of values.

(a) $u = 7$, $v = 1$, $w = 3$

(b) $u = 0.75$, $v = 0.57$, $w = 0.15$

(c) $u = 32$, $v = 8$, $w = 96$

(d) $u = 3500$, $v = 500$, $w = -2500$

(e) $u = 2.12$, $v = 0.62$, $w = 2.88$

3 Use the given instructions to create formulae.

(a) To calculate your Body Mass Index (B) you need to know your height in metres (h) and your weight in kilograms (w). Square your height, then divide your weight by the result.

(b) To work out the surface area (S) of a cuboid, you need its width (w), length (l) and height (h). Multiply the length by the width, then multiply the width by the height, then the height by the length. Add the three results together, then double the answer.

Rearranging formulae

> Rearranging formulae means getting the subject on one side of the equals sign and everything else on the other side of the equals sign.

> It is also known as changing the subject, or transforming formulae.

A Changing the subject – types 1 and 2

>> **key fact** Changing the subject of a formula means rearranging a formula to get one letter on its own and all of the other letters on to the other side of the equals sign.

It is also known as **transformation** of formulae.

Type 1: When x is not 'bound' up with anything else:

$$x + a = b$$
$$x + a - a = b - a \qquad \textit{here subtract a from both sides}$$
$$x = b - a$$

Type 2: When x is 'bound' in a multiplication:

$$xy = a$$
$$\frac{xy}{y} = \frac{a}{y} \qquad \textit{divide both sides by y}$$
$$x = \frac{a}{y}$$

Look at this formula: $2x^2 y = z$

$$x^2 = \frac{z}{2y} \qquad \textit{divide both sides by 2y}$$
$$x = \sqrt{\frac{z}{2y}} \qquad \textit{remove the square by finding the square root of both sides}$$

B Combinations of types 1 and 2

1. Make x the subject of this formula:

$$q + 6x = p$$
$$q + 6x - q = p - q \qquad \textit{isolate the xs first by taking q from both sides}$$
$$6x = p - q$$

2. We need only one x on the left-hand side, so we 'undo' $6x$ by dividing each side by 6:

$$x = \frac{p - q}{6}$$

C   What to do if the subject has a minus sign

1 Sometimes you may arrange a formula only to find that the left-hand side contains the subject with a minus sign in front of it, for example $-A = sx - 2t$.

>> **key fact**   **If this happens, multiply both sides of the equation by -1.**

This has the effect of changing the sign of **every term** in the formula, because: $-A \times -1 = A$.

If $-A = sx - 2t$, then $A = -sx + 2t$

2 Make r the subject of $3M - 2r = 4N$.

$-2r$	$= 4N - 3M$	*subtract $3M$ from both sides*
$-r$	$= 2N - \frac{3M}{2}$	*divide both sides by 2*
r	$= -2N + \frac{3M}{2}$	*multiply both sides by -1*
r	$= \frac{3M}{2} - 2N$	*rearrange right-hand side for neatness*

D   When factorisation is required

1 If the letter you want to make the subject of a formula occurs in **two terms**, you may need to **collect** these terms and then **factorise** them.

2 Make x the subject of $2x = px + q$.

$2x - px$	$= q$	*subtract px from both sides*
$(2 - p)x$	$= q$	*factorise the left hand side*
x	$= \frac{q}{2-p}$	*divide both sides by $2 - p$*

>> practice questions

In these questions, make x the subject of the formula.

1   $x + 9 = r$

2   $x - z = a$

3   $5x + 4y = 16$

4   $mx^2 = c$

5   $\frac{1}{4}x = m$

6   $\frac{x}{p} = p + c$

7   $\frac{mx}{b} = c$

8   $5 - fx = 3x + p$

In these questions, make y the subject of the formula.

9   $2x + 5y = 9$

10   $x - 2y = 10$

Using brackets in algebra

Expanding brackets means multiplying terms to remove brackets from equations or expressions.

To expand an expression, you need to multiply each term inside the bracket by the term outside the bracket.

A Expanding brackets

>> **key fact** To expand brackets, multiply everything in the bracket by the term outside the bracket.

① Expand $3(2a + b) = 6a + 3b$

② Expand $6(y + 3) = 6y + 18$

It is a common mistake to forget to multiply the **second** term.
If you worked out $6y + 3$ for the answer to the second equation, you'd be wrong!

③ Expand $3(2x + 4) = 6x + 12$

This example uses the fact that $3 \times 2x = 6x$.

④ Expand $x(4x + 9) = 4x^2 + 9x$

Notice here that x is the term outside the bracket and that $x \times x = x^2$.

⑤ Expand $5x(3x - 2) = 15x^2 - 10x$

Notice that $5x \times 3x = 5 \times 3 \times x \times x = 15x^2$.

B Factorising expressions

>> **key fact** Factorising is the opposite of expanding brackets.
Find the highest common factor (HCF) of all of the terms in the expression you are trying to factorise. The HCF must appear outside the brackets.

Factorise $4x^2 + 8x$.

Here the HCF is the largest term that is a factor of $4x^2$ and $8x$. The HCF must be $4x$.

So we now have $4x^2 + 8x = 4x(? + ?)$.

Ask yourself, 'What do I multiply $4x$ by, to make $4x^2$, and what do I multiply $4x$ by to make $8x$?'

So the final answer $= 4x(x + 2)$.

C Solving equations with brackets

Method 1

Solve $4(x + 3) = 28$.

Expand the brackets:

$$4x + 12 = 28$$

Solve it like any other equation:

$$4x + 12 = 28$$
$$4x = 16$$
$$x = 4$$

Method 2

Solve $4(x + 3) = 28$.

$4(x + 3)$ is a product, it is $4 \times (x + 3)$.

So divide by 4 to undo the multiplication:

$$x + 3 = 7$$
$$x = 4$$

D Solving equations with two sets of brackets

Solve $3(x + 4) + 5(x - 6) = 5x - 3$.

Step 1: Expand the brackets and simplify.

$$3x + 12 + 5x - 30 = 5x - 3$$
$$8x - 18 = 5x - 3$$

Step 2: Solve the equation. You should find that:

$$x = 5$$

>> practice questions

1. $2(x + 5) = 18$

2. $5(x - 2) = 40$

3. $4(3x - 7) - 5(2x - 4) = 3x$

4. $2(7x - 7) - 2(4x + 5) = 21 - 4x$

5. $5 + (7x - 7) - 2(4x + 5) = 2x + 4$

6. $6(4x - 7) + 3(13 - 3x) = 20x - 23$

7. $8(2x - 9) - 2(9 - 2x) = 14x + 3$

8. **Factorise:**

    (a) $14x^2 + 7x$

    (b) $36y^2 - 9y$

9. **Factorise:**

    (a) $15y^4 + 25y^2$

    (b) $100a^2 + 20ab^3$

Factorising quadratic expressions

> Quadratic expressions are expressions whose highest power is 2. In general, they have the form $ax^2 + bx + c$.
>
> A difference of two squares is an expression of the form $A^2 - B^2$.

A Quadratics without a number term

1. The terms in an expression of the form $ax^2 + bx$ always have a common factor containing x.

2. *Example*: Factorise $4x^2 + 12x$.

   x is obviously a common factor.

   Also, the highest common factor of 4 and 12 is 4.

   Therefore extract the factor $4x$.

   $4x^2 + 12x = 4x(x + 3)$.

B Quadratics of the form $x^2 + bx + c$

1. Expressions of this type factorise to two brackets: $(x + p)(x + q)$.

   In this case, $pq = c$ and $p + q = b$, so look for two numbers that multiply to make c and add to make b.

2. *Example*: Factorise $x^2 - 9x + 14$.

   You are searching for two numbers that multiply to make 14 and add to make -9.

   They must be -2 and -7.

   Now rewrite the bx term using these numbers: $-2x$ and $-7x$.

   So $x^2 - 9x + 14 = x^2 - 2x - 7x + 14$
   $$= x(x - 2) - 7(x - 2)$$
   $$= (x - 7)(x - 2).$$

C The general quadratic

1. The general quadratic expression is of the form $ax^2 + bx + c$, where a, b and c can take any integer value. To factorise this type of expression, follow these steps:

   * Multiply a by c.

   * Look for factors of this number that add to make b, as in the last section.

2. *Example*: Factorise $6x^2 + 7x - 5$.

   First multiply 6 by -5 to give -30.

   You are searching for two numbers that multiply to make -30 and add to make 7.

   These are -3 and 10.

   Now rewrite the bx term using these numbers: $-3x$ and $10x$.

   So $6x^2 + 7x - 5 = 6x^2 - 3x + 10x - 5$
   $$= 3x(2x - 1) + 5(2x - 1)$$
   $$= (3x + 5)(2x - 1).$$

D Difference of two squares

1 This is a special factorisation you need to recognise.

Write the expression in the form $A^2 - B^2$. This factorises to $(A + B)(A - B)$.

2 *Example*: Factorise $x^2 - 4$.

x^2 is the square of x and 4 is the square of 2.

So $x^2 - 4 = x^2 - 2^2$. In the notation above, $A = x$ and $B = 2$.

So $x^2 - 4 = (x - 2)(x + 2)$.

E Quadratics with an extra common factor

1 Questions on factorisation often begin 'Factorise completely…' You should always check for a common factor before starting the processes described above.

2 *Example*: Factorise $4x^2 + 20x + 24$.

The common factor is 4: extract this first.

$= 4(x^2 + 5x + 6)$

Now factorise $x^2 + 5x + 6$ to get $(x + 2)(x + 3)$.

The complete factorisation is $4x^2 + 20x + 24 = 4(x + 2)(x + 3)$.

If you forget to extract the common factor first, you could end up with any of these:

$(4x + 8)(x + 3)$ $(2x + 4)(2x + 6)$
$(x + 2)(4x + 12)$

These are all correct factorisations, but the brackets contain common factors of 4 or 2 which need to be extracted to complete the question.

3 *Example*: Factorise $48p^2 - 75$.

This looks a bit like a difference of two squares, but the numbers aren't squares.

Extract the common factor, 3.

$48p^2 - 75 = 3(16p^2 - 25q^2) = 3(4p - 5q)(4p + 5q)$.

4 If the coefficient of x^2 is -1, treat this as a common factor of -1.

Example: Factorise $3x - 2 - x^2$.

First rearrange the terms into the correct order:

$3x - 2 - x^2 = -x^2 + 3x - 2$

Extract the common factor -1:

$-x^2 + 3x - 2 = -1(x^2 - 3x + 2)$
$= -1(x + 2)(x - 1)$.

The -1 can now be 'hidden' by incorporating it into the second bracket:

$1(x + 2)(x - 1) = (x + 2)(1 - x)$.

5 If the coefficient of x^2 is any other negative number, the common factor will be negative:

$-20x^2 - 10x + 60 = -10(2x^2 + x - 6)$
$= -10(x + 2)(2x - 3)$
$= 10(x + 2)(3 - 2x)$.

>> practice questions

Factorise the following quadratic expressions:

1 $x^2 + 5x$

2 $4x^2 - 12x$

3 $x^2 + 6x + 8$

4 $t^2 - 5t - 14$

5 $25x^2 - 16$

6 $2x^2 + 5x - 12$

7 $4h^2 - 4h - 3$

8 $8x^2 - 14x - 15$

9 $3x^2 + 15x + 12$

10 $6n^2 + 18n - 108$

11 $20k^2 - 45$

12 $12x^2 + 22x - 70$

Algebraic fractions

$\frac{3x^2}{6x}$ and $\frac{8(x+9)}{6x}$ are examples of algebraic fractions.

Fractions can only be simplified by cancelling, when and only when there is a common factor in the numerator and the denominator.

A Simplifying fractions

1 Simplify $\frac{3x^4}{6x}$

$\frac{3x^4}{6x} = \frac{3 \times x \times x \times x \times x}{6 \times x} = \frac{1 \times x \times x \times x}{2} = \frac{x^3}{2}$

2 Simplify $\frac{8(x+9)}{2x}$

Look for the common factors. Obviously 8 and 2 will cancel because they have common factors, so:

$\frac{8(x+9)}{2x} = \frac{4(x+9)}{x}$

>> key fact You cannot divide the xs in example 2, because x is not a common factor.

B The number line

>> key fact To multiply algebraic fractions:

Look for and cancel any factor that is common to the numerator and the denominator.

Only when it is not possible to cancel any further, multiply the remaining numerators and denominators.

Simplify $\frac{x^6}{y^2} \times \frac{x^2 y}{z} \times \frac{z^2}{x^3}$

Cancel the xs first: x^6 and x^3 will cancel.

Then cancel the y^2 with the y, leaving y in the denominator.

Now cancel z^2 with z

So $\frac{x^6}{y^2} \times \frac{x^2 y}{z} \times \frac{z^2}{x^3} = \frac{x^3}{y} \times x^2 \times z$

$= \frac{x^5 z}{y}$

C Dividing algebraic fractions

>> **key fact** When you divide by an algebraic fraction, it is the same as multiplying by its reciprocal.

1 Look at what happens when you work in numbers.

$4 \div \frac{1}{2}$ means how many $\frac{1}{2}$s are there in 4?

Clearly there are eight halves in 4, so $4 \div \frac{1}{2} = 8$.

But the reciprocal of $\frac{1}{2}$ is 2. $8 = 4 \times 2$.

>> **key fact** The reciprocal of a number is the number you multiply by to get an answer of 1.

2 $4y \div \frac{x}{z} = \frac{4y}{1} \times \frac{z}{x}$

$= \frac{4yz}{x}$

D Adding and subtracting algebraic fractions

>> **key fact** The lowest common denominator is the LCM of the denominators of the original fractions.

As with ordinary fractions, you can't add or subtract unless the fractions have a **common denominator**.

Simplify $\frac{3x + 1}{2} + \frac{2x - 3}{3}$

The denominators are 2 and 3, so the lowest common denominator (LCD) is 6.

Multiplying the top and bottom of the first fraction by 3, $\frac{3x + 1}{2} = \frac{3(3x + 1)}{6}$.

Multiplying the top and bottom of the second fraction by 2, $\frac{2x - 3}{3} = \frac{2(2x - 3)}{6}$.

So $\frac{3x + 1}{2} + \frac{2x - 3}{3} = \frac{3(3x + 1)}{6} + \frac{2(2x - 3)}{6}$.

$= \frac{9x + 3}{6} + \frac{4x - 6}{6}$

$= \frac{9x + 3 + 4x - 6}{6} = \frac{13x - 3}{6}$

>> practice questions

Simplify the following:

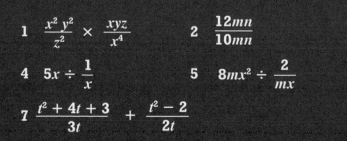

1 $\frac{x^2 y^2}{z^2} \times \frac{xyz}{x^4}$

2 $\frac{12mn}{10mn}$

3 $\frac{3(x + 4)}{6(x + 4)(x - 4)}$

4 $5x \div \frac{1}{x}$

5 $8mx^2 \div \frac{2}{mx}$

6 $\frac{2a + 3}{4} - \frac{a + 2}{5}$

7 $\frac{t^2 + 4t + 3}{3t} + \frac{t^2 - 2}{2t}$

Solving equations

 An equation is a mathematical puzzle that has a solution. Finding the solution is called solving the equation.

 Always perform the same operation on both sides of an equation.

 To 'undo' the effect of an operation, apply the inverse (opposite) operation.

A Linear equations

A **linear** equation doesn't contain any powers of the unknown letter.

>> **key fact** You must always perform the same operation on both sides of an equation.

Examples:

$3x + 4 = 19$

$3x = 15$	*subtract 4*
$x = 5$	*divide by 3*

$5 - 6n = 3$

$-6n = -2$	*subtract 5*
$-n = -\frac{1}{3}$	*divide by 6*
$n = \frac{1}{3}$	*multiply by -1*

Sometimes the unknown letter occurs more than once in the equation:

$4x - 6 = 2x + 3$

$2x - 6 = 3$	*subtract 2x*
$2x = 9$	*add 6*
$x = 4.5$	*divide by 2*

B Equations with brackets

To solve $3(5p - 4) = 21$ you can use one of two methods:

Divide by the number outside the bracket ...

$5p - 4 = 7$	
$5p = 11$	*add 4*
$p = 2.2$ or $\frac{11}{5}$	*divide by 5*

or expand the bracket

$15p - 12 = 21$	
$15p = 33$	*subtract 5*
$p = \frac{33}{15}$	*divide by 15*
$p = \frac{11}{5}$ or **2.2**	*simplify*

This equation doesn't seem to contain a bracket, but the division acts like one:

$$\frac{2R + 3}{7} = 1$$

$$2R + 3 = 7 \qquad \textit{multiply by 7}$$

$$2R = 4 \qquad \textit{subtract 3}$$

$$\boldsymbol{R = 2} \qquad \textit{divide by 2}$$

C More complex equations

1 This equation needs to be expanded and simplified:

$$7(2x + 3) - 3(3x - 1) = 9$$

$$14x + 21 - 9x + 3 = 9 \qquad \textit{expand}$$

$$5x + 24 = 9 \qquad \textit{simplify}$$

$$5x = -15 \quad \textit{subtract 24}$$

$$\boldsymbol{x = -3} \quad \textit{divide by 5}$$

2 This equation needs to be multiplied by a common denominator:

$$\frac{w - 3}{2} + \frac{2w}{3} = 2$$

$$3(w - 3) + 2(2w) = 12 \qquad \textit{multiply by 6}$$

$$3w - 9 + 4w = 12 \qquad \textit{expand}$$

$$7w - 9 = 12 \qquad \textit{expand}$$

$$7w = 21 \qquad \textit{simplify}$$

$$\boldsymbol{w = 3} \qquad \textit{divide by 7}$$

3 You can use inversion to solve this equation:

$$\frac{5}{x} + \frac{3}{x} = \frac{1}{2}$$

$$\frac{8}{x} = \frac{1}{2} \qquad \textit{add the fractions}$$

$$\frac{x}{8} = \frac{2}{1} \qquad \textit{invert (take reciprocals)}$$

$$\boldsymbol{x = 16} \qquad \textit{multiply by 8}$$

D Simple non-linear equations

1 Other types of equation can be solved in the same way as linear ones, as long as they are not too complicated.

2 Examples:

$$3x^3 - 6 = 18$$

$$3x^3 = 24 \qquad \textit{add 6}$$

$$x^3 = 8 \qquad \textit{divide by 3}$$

$$\boldsymbol{x = 2} \qquad \textit{cube root}$$

$$\frac{4}{\sqrt{a}} = 3$$

$$4 = 3\sqrt{a} \qquad \textit{multiply by } \sqrt{a}$$

$$\frac{4}{3} = \sqrt{a} \qquad \textit{divide by 3}$$

$$\frac{16}{9} = \boldsymbol{a} \qquad \textit{square}$$

>> practice questions

Solve these equations.

1 $4x + 7 = 35$

2 $10 - 3k = -5$

3 $2b + 5 = 7b - 13$

4 $4(2x - 1) = 20$

5 $\frac{5u - 1}{3} = 8$

6 $2(10y + 3) + 4(5 - 3y) = 0$

7 $\frac{2N - 1}{2} + \frac{N}{5} = 1$

8 $\frac{8}{z} + \frac{2}{z} = 5$

9 $2x^2 - 5 = 7.5$

10 $\frac{2}{\sqrt[5]{x}} = 1$

Equations of proportionality

 Linked quantities may conform to a number of relationships: for example, direct proportion, where one quantity is a multiple of another, or inverse proportion, where the product of the quantities is fixed.

 One quantity may be proportional or inversely proportional to a power of the other.

A Direct proportion

1 Suppose that two quantities y and x are in direct proportion.

>> **key fact** 'y is proportional to x' is written $y \propto x$.

If you plot them against each other on a graph, they form a straight line through the origin.

>> **key fact** If $y \propto x$, then $y = kx$ for some number k, called the constant of proportionality.

2 If two quantities are in direct proportion, and you know a pair of corresponding values, you can find the constant of proportionality and establish the equation linking them.

y is proportional to x. When $x = 250$, $y = 6.25$. What is the equation of proportionality?

$y \propto x$, so $y = kx$.

Substituting the known values gives $6.25 = k \times 250$, so $k = \frac{6.25}{250} = 0.025$ or $\frac{1}{40}$.

The equation of proportionality is therefore $y = 0.025x$ or $y = \frac{x}{40}$.

3 Note that other linear equations with a constant term (e.g. $y = 2x + 5$) do not describe proportionality. This is because, for example, doubling x does not result in a doubling of y.

B Inverse proportion

1 Two quantities are in inverse proportion if, when one is multiplied by a number, the other is divided by that number (e.g. if one is doubled, the other is halved).

>> **key fact** A relationship of this type is written $y \propto \frac{1}{x}$.

2 The equation of inverse proportionality is of the form $y = \frac{k}{x}$.

3 y is inversely proportional to x. When $x = 30$, $y = 0.6$.

Find the equation of proportionality, and the value of y when $x = 6$. $y \propto \frac{1}{x}$ so $y = \frac{k}{x}$.

Substituting the known values gives $30 = \frac{k}{0.6}$, so $k = 30 \times 0.6 = 18$.

The equation of proportionality is therefore $y = \frac{18}{x}$.

When $x = 6$, $y = \frac{18}{x} = \frac{18}{6} = 3$.

C Other types of proportionality

There are other types of proportionality involving other powers of x:

y is proportional to x	$y \propto x$	$y = kx$
y is inversely proportional to x	$y \propto \frac{1}{x}$	$y = \frac{k}{x}$
y is proportional to x^2	$y \propto x^2$	$y = kx^2$
y is inversely proportional to x^2	$y \propto \frac{1}{x^2}$	$y = \frac{k}{x^2}$
y is proportional to x^3	$y \propto x^3$	$y = kx^3$

The procedure for dealing with these relationships is exactly as for direct and inverse proportion: substitute known values, find k, then write down the equation of proportionality. This can then be used to answer any additional questions.

y is proportional to x^2. When $x = 4$, $y = 200$. What is the equation of proportionality, and what is the value of y when $x = 2$?

$y \propto x^2$, so $y = kx^2$.

Substituting the known values gives $200 = k \times 4^2$, so $16k = 200$ and $k = 12.5$.

The equation of proportionality is therefore $y = 12.5x^2$.

When $x = 2$, $y = 12.5x^2 = 12.5 \times 2^2 = 12.5 \times 4 = 50$.

1 **The braking distance (y metres) of a car is proportional to the square of its speed (x km/h).**
   Find the equation connecting x and y, and complete the table.

speed (x km/h)			64	80
braking distance (y metres)		10	24	

2 **The distances of the planets from the Sun are often measured in astronomical units (au). 1 au is defined to be the distance of the Earth from the Sun (about 150 million km).**

   The speed (v km/sec) of a planet in its orbit is inversely proportional to the square root of its orbital distance (d au) from the Sun.

   Find the rule linking d and v, and complete the table.

planet	Mercury	Earth	Neptune
orbital distance (d au)		1	30
speed (v km/s)	48		5

Trial and improvement

Sometimes you can't find an exact solution to an equation, but can find a reasonable approximation using trial and improvement.

The accuracy of the approximation depends on the number of trials: the more trials you do, the closer you can get to the solution.

A Decimal searches

Example:

Solve the equation $x^2 = 3$, correct to 2 decimal places.

This is a very simple equation which you could solve easily with a press of the square root key! However, it demonstrates the solution process clearly. Notice how the search concentrates on whole numbers, then decimals with one decimal place, then two.

>> **key fact** In a decimal search, the solution is found one significant figure at a time.

Record the results of the trials, together with your decisions, in a table like the one opposite:

Therefore the solution is $x = 1.73$, to 2 dp.

You have to test $x = 1.735$, because you need to know whether x is closer to 1.73 or 1.74. It's closer to 1.73, so whatever the *exact* solution of the equation is, when rounded to 2 dp, it's 1.73.

>> **key fact** You try out likely solutions in an equation to see how closely they fit. You use the results to make better guesses. That's why it's 'trial and improvement', not just 'trial and error'.

x	x^2	Comments
1	1	Too small, so $x > 1$. Try $x = 2$.
2	4	Too big, so $x < 2$. x is between 1 and 2. Move on to numbers with 1 decimal place: try 1.5 first, as it's halfway between 1 and 2.
1.5	2.25	Too small, so $x > 1.5$. Try 1.7, as it's about halfway between 1.5 and 2.
1.7	2.89	Too small, so $x > 1.7$. Try 1.8.
1.8	3.24	Too big, so $x < 1.8$. x is between 1.7 and 1.8. Move on to numbers with 2 decimal places. Try 1.75, as it's halfway between 1.7 and 1.8.
1.75	3.0625	Too big, so $x < 1.75$. Try 1.72, as it's about halfway between 1.7 and 1.75.
1.72	2.9584	Too small, so $x > 1.72$. Try 1.73.
1.73	2.9929	Too small, so $x > 1.73$. Try 1.74.
1.74	3.0276	Too big, so $x < 1.74$. x is between 1.73 and 1.74, so the answer is one of these. You now need to know whether it's closer to 1.73 or 1.74. Trying 1.735 will decide.
1.735	3.010225	Too big, so $x < 1.735$. x is between 1.73 and 1.735.

B Rearranging the equation

Example:

Solve the equation $p^3 - 10 = p$, correct to 1 decimal place.

As you try different values for p, both sides of the equation change.

It's much easier to 'hit the target' if you're aiming for a fixed number, so rearrange the equation to read $p^3 - p = 10$.

p	$p^3 - p$	Comments
1	0	Too small, so $p > 1$. Try $p = 2$.
2	6	Too small, so $p > 2$. Try $p = 3$.
3	24	Too big, so $p < 3$. p is between 2 and 3. Move on to numbers with 1 decimal place: try 2.5 first, as it's halfway between 2 and 3.
2.5	13.125	Too big, so $p < 2.5$. Try 2.3, as it's about halfway between 2 and 2.5.
2.3	9.867	Too small, so $p > 2.3$. Try 2.4.
2.4	11.424	Too big, so $p < 2.4$. p is between 2.3 and 2.4, so the answer is one of these. You now need to know whether it's closer to 2.3 or 2.4. Trying 2.35 will decide.
2.35	10.627875	Too big, so $p < 2.35$. p is between 2.3 and 2.35.

Therefore the solution is $p = 2.3$, to 1 dp.

As you might have guessed, the solution ($p = 2.3089...$) is very close to 2.3!

C Exact answers

Sometimes you may locate an exact answer by trial and improvement. Suppose you had to solve

$$\frac{552.96}{x^3} = 5.$$

By trying whole numbers, you would find that x is between 4 and 5.

Looking between 4 and 5, you would eventually try $x = 4.8$. This is exactly right, so no rounding of the answer is necessary.

>> practice questions

Use trial and improvement to find solutions to the following equations.
Give answers correct to 1 decimal place.

1 $x^3 - 2x^2 = 1$
2 $x^3 + 4x - 18 = 0$
3 $5y^2 - 17y = 1.75$
4 $6m^2 - 9 = 6m$
5 $5t + 1.5 = 2t^3$
6 $5x^3 = 60.835$

Quadratic equations

A quadratic equation is one that can be written in the form $ax^2 + bx + c = 0$. b or c could be 0, but a cannot.

Solve quadratic equations by factorising, completing the square, using the quadratic formula, trial and improvement (see page 34), or using a graph (see page 48).

Quadratic equations may have 2 roots (solutions), 1 repeated root or no roots.

Quadratic equations may be 'hidden' in equations using fractions, for example.

A Equations that factorise

1 To solve $x^2 + 2x = 24$, follow these steps:

$x^2 + 2x - 24 = 0$ *subtract 24 from both sides*

The equation is now in the form

$ax^2 + bx + c = 0$.

$(x + 6)(x - 4) = 0$ *factorise the left-hand side*

So $(x + 6) = 0$ or $(x - 4) = 0$.

>> **key fact** If the product of two brackets is zero, one of them must be zero.

So $x = -6$ or $x = 4$.   *these are the values of x that make either bracket zero*

2 Sometimes you may find the equation has a repeated root (also known as a **double root**).

Example: Solve $3x^2 - 6x + 3 = 0$.

$x^2 - 2x + 1 = 0$ *divide by the common factor, 3*

$(x - 1)^2 = 0$ *factorise*

So $(x - 1) = 0$ and therefore $x = 1$ (repeated).

B Completing the square

1 You are expected to be able to complete the square for expressions of the form $x^2 + bx + c = 0$.

>> **key fact** Completing the square involves creating a squared bracket that matches the coefficients in $x^2 + bx$. This is always $\left(x + \dfrac{b}{2} \right)^2$.

Example: Solve the equation $x^2 + 4x - 10 = 0$.

You can create $x^2 + 4x$ by squaring $(x + 2)$.

$(x + 2)^2 = x^2 + 4x + 4$

So $x^2 + 4x = (x + 2)^2 - 4$.

So $x^2 + 4x - 10 = [(x + 2)^2 - 4] - 10$
$= (x + 2)^2 - 14$.

The equation is now $(x + 2)^2 - 14 = 0$.

So $(x + 2)^2 = 14$ *add 14 to both sides*

So $x + 2 = \pm\sqrt{14}$ *square root*

So $x = -2 \pm \sqrt{14}$ *subtract 2*

Correct to 2 dp, the roots are 1.74 and -5.74.

Sometimes the equation has no roots.

Example: Solve the equation $x^2 - 2x + 5 = 0$.

$(x - 1)^2 = x^2 - 2x + 1$

So $x^2 - 2x = (x - 1)^2 - 1$.

So $x^2 - 2x + 5 = [(x - 1)^2 - 1] + 5$
$= (x - 1)^2 + 4$.

The equation is now $(x - 1)^2 + 4 = 0$.

So $(x - 1)^2 = -4$ *subtract 4*

No real number when squared gives -4, so the equation has no solutions.

C The quadratic formula

When completing the square is applied to the general quadratic equation $ax^2 + bx + c = 0$, the solutions are given by the **quadratic formula**.

>> **key fact** If $ax^2 + bx + c = 0$, $x = \dfrac{-b \pm \sqrt{b^2 - 4ac}}{2a}$.

If the quantity inside the square root is positive, there are two roots; if it is zero, there is a repeated root; if it is negative, there are no roots.

Example: Solve $3x^2 + 2x - 1 = 0$.

$a = 3, b = 2, c = -1$

$x = \dfrac{-b \pm \sqrt{b^2 - 4ac}}{2a} = \dfrac{-2 \pm \sqrt{2^2 - 4 \times 3 \times (-1)}}{2 \times 3} = \dfrac{-2 \pm \sqrt{4 + 12}}{6}$

$= \dfrac{-2 \pm \sqrt{16}}{6} = \dfrac{-2 \pm 4}{6}$

$= \mathbf{-1}$ **or** $\dfrac{1}{3}$

You do not need to remember the formula: it will be provided on the exam paper.

D 'Hidden' quadratic equations

Sometimes quadratic equations can be generated by other kinds of equation, often involving fractions.

Example: Solve $2x - \dfrac{5}{x} = 9$.

$2x^2 - 5 = 9x$ *multiply all terms by x*

$2x^2 - 9x - 5 = 0$ *subtract 9x*

$(2x + 1)(x - 5) = 0$ *factorise*

So $\boldsymbol{x = -\frac{1}{2}}$ or $\boldsymbol{x = 5}$

>> practice questions

Solve the following equations. If no method is given, use any one you like.

1   $x^2 - 5x = 6$ (factorise)

2   $x^2 - 4x - 7 = 0$ (complete the square)

3   $3x^2 + x - 7 = 0$ (use the formula)

4   $2x^2 + 3x - 5 = 0$

5   $5x - \dfrac{2}{x} = 4$

6   $\dfrac{2}{x - 2} + \dfrac{3}{x + 1} = 2$

Functions

✂ **A function is a way of describing a sequence of mathematical operations.**

✂ **Most functions have inverses or opposites.**

A Function notation

1 A **function** is a way of describing a sequence of mathematical operations. These can then be applied to a set of numbers. Functions are also known as **mappings**.

2 Suppose you wanted to describe the operations 'multiply by 2, then add 5' as a function called f.

This can be written $f(x) = 2x + 5$ pronounced 'f of x equals $2x + 5$',

or $f : x \mapsto 2x + 5$ pronounced 'f maps x to $2x + 5$'.

Either of these completely describes what the function does.

3 The notation can also be used to describe what happens to particular numbers when f is used on them. For example, if f is used on the number 3,

$f(3) = 11$

or $f : 3 \mapsto 11$

You can also say that if $x = 3$, $f(x) = 11$, or that if 3 is the **input** for f, 11 is the **output**.

4 Functions are also sometimes illustrated with diagrams:

INPUT x ×2 +5 $2x + 5$ OUTPUT

B Types of function

In the following list, letters such as a, b, c, etc. stand for fixed numbers or **constants**. You need to be familiar with several different types of function:

type	function definition	example	page
linear	$f(x) = ax + b$	$f(x) = 2x - 1$	46, 40, 52
quadratic	$f(x) = ax^2 + bx + c$	$f(x) = x^2 - 2x + 1$	48, 50, 54
cubic	$f(x) = ax^3 + bx^2 + cx + d$	$f(x) = x^3 - 10$	48, 50
reciprocal	$f(x) = \frac{a}{x}$	$f(x) = \frac{1}{x}$	48, 50
exponential	$f(x) = a^x$	$f(x) = 2^x$	
trigonometric	$f(x) = a \sin bx$ $f(x) = a \cos bx$ $f(x) = a \tan bx$	$f(x) = \sin x$ $f(x) = \frac{1}{2} \cos x$ $f(x) = \tan 2x$	72, 50

C Applications of functions

1 The position-to-term rule for a sequence is an example of a function.

This will be given in the form $u_n = f(n)$. For example, this mapping diagram shows powers of 2, with $f(n) = 2^n$ (an exponential function).

2 Functions can be used to define graphs.

If $g(x) = x^2$, the graph $y = g(x)$ is the graph of $y = x^2$.

D Equations from functions

1 Suppose $h(x) = x^2 - 3$. Then $h(x) = 6$ is an equation that can be solved.

Replace $h(x)$ with its definition, then proceed as normal:

$h(x) = 6$, so $x^2 - 3 = 6$.

$$x^2 = 9 \qquad \text{add 3 to both sides}$$
$$x = \pm 3 \qquad \text{square root}$$

2 Suppose $f(x) = x^2$ and $g(x) = 2x - 1$. Then $f(x) = g(x)$ is an equation.

$f(x) = g(x)$, so $x^2 = 2x - 1$.

$$x^2 - 2x + 1 = 0 \qquad \text{subtract } (2x - 1) \text{ from both sides}$$
$$(x - 1)^2 = 0 \qquad \text{factorise}$$
$$x - 1 = 0 \qquad \text{square root}$$
$$x = 1 \qquad \text{add 1}$$

E Change of variable

Suppose $f(x) = x^2 + x + 1$.

Instead of using a number as input, you can use another algebraic expression.

For example,
$$f(x + 1) = (x + 1)^2 + (x + 1) + 1$$
$$= (x^2 + 2x + 1) + (x + 1) + 1$$
$$= x^2 + 3x + 3.$$
$$f(2x) = (2x)^2 + 2x + 1$$
$$= 4x^2 + 2x + 1.$$

>> practice questions

In these questions,

$f(x) = 2x - 1$, $g(x) = x^2$, $h(x) = \dfrac{1}{1 + x}$.

1 Evaluate:

(a) $f(3)$ (b) $f(-5)$ (c) $f(0.1)$

(d) $g(4)$ (e) $g(0)$ (f) $h(-3)$

2 Find expressions to represent:

(a) $f(x + 1)$ (b) $f(2x)$ (c) $f(x^2)$

(d) $g(x - 2)$ (e) $g\left(\dfrac{1}{x}\right)$ (f) $h(x - 1)$

(g) $h\left(\dfrac{x}{2}\right)$ (h) $f(f(x))$

Inequalities and regions

- An inequality describes a range of numbers.
- Inequalities can be solved in the same way as equations.
- Inequalities can be used to represent regions on a co-ordinate grid.

A Linear inequalities

To solve a linear inequality, follow the same procedures as for linear equations.

Example: Solve $4x + 5 < 21$.

$4x < 16$	*subtract 5 from both sides*
$x < 4$	*divide by 4*

The solution to an inequality is a range of numbers. This can be illustrated on a number line. The open circle shows that x is **less than** 4 (the value 4 itself is *not* included).

Example: Solve $10 + x \leqslant 4 + 3x$.

$10 - 2x \leqslant 4$	*subtract 3x from both sides*
$-2x \leqslant -6$	*subtract 10*
$-x \leqslant -3$	*divide by 2*
$x \leqslant 3$	*multiply by -1*

>> **key fact** When you multiply or divide by a negative number, the inequality changes direction.

The solid circle shows that x is **greater than or equal to** 3 (the value 3 is included).

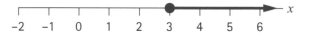

$-1 < 3x + 2 \leqslant 11$ means that $3x + 2$ is between -1 and 11; 11 is an allowed value but -1 is not. To solve it, simply use the same operations on all three parts of the inequality:

$-3 < 3x \leqslant 9$	*subtract 2 from both sides*
$-1 < x \leqslant 3$	*divide by 3*

This number line shows the result, a **finite** range of values.

B Quadratic inequalities

>> **key fact** To solve a quadratic inequality, first find the roots of the corresponding equation.

Example: Solve $x^2 + 3x > 10$.

First rearrange to give $x^2 + 3x - 10 \geqslant 0$, then use the equation $x^2 + 3x - 10 = 0$. This factorises to give $(x + 5)(x - 2) = 0$. The **critical values** of x, where the quadratic expression changes from positive to negative, are therefore -5 and 2. There are two ways you can solve the inequality:

Method 1: examine signs of brackets

range of x	signs of brackets	sign of product
$x < -5$	$(x + 5)$ negative, $(x - 2)$ negative	$(x + 5)(x - 2)$ positive ✓
$-5 < x < 2$	$(x + 5)$ positive, $(x - 2)$ negative	$(x + 5)(x - 2)$ negative ✗
$x > 2$	$(x + 5)$ positive, $(x - 2)$ positive	$(x + 5)(x - 2)$ positive ✓

So $x < -5$ or $x > 2$.

Method 2: draw a sketch graph

The highlighted parts of the curve are above the axis, so the solution is $x < -5$ or $x > 2$.

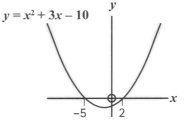
$y = x^2 + 3x - 10$

C Regions

Every straight line on a co-ordinate grid divides the grid into two parts called **regions**. These regions can be described using inequalities.

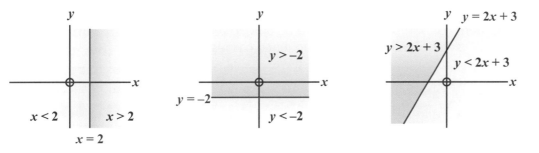

For vertical lines of the form $x = a$, the region to the left is $x < a$; the region to the right is $x > a$.

For all other lines, the rule is to use $y > ...$ for the region above and $y < ...$ for the region below. To include points on the line itself, use $\leqslant$ or $\geqslant$.

Example: Draw, and find the area of, the triangle described by the inequalities $x > -2$, $y > 2$ and $x + 2y < 8$.

The first two inequalities are easy to draw, but the last one needs to be rearranged to make y the subject: $2y < 8 - x$, so $y < 4 - \frac{1}{2}x$. Plot these on a grid.

The base of the triangle is 6 units long and the height is 3 units.

The area is $\frac{1}{2} \times 6 \times 3 = 9$ square units.

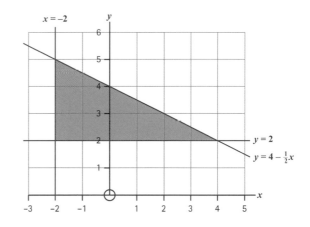

>> practice questions

1 Solve the following inequalities.

   (a) $2x + 1 \leqslant 5$ (b) $7x + 3 > 5x - 4$

   (c) $-3h < -12$ (d) $-3 < \frac{t}{3} + 2 \leqslant 5$

   (e) $x^2 \geqslant 36$ (f) $x^2 + x < 20$

   (g) $x^2 > 1 - 4x$

2 Draw a co-ordinate grid with x- and y-axes from -6 to 6.
   Draw the triangle described by these inequalities: $y < x + 4$; $y < 6 - x$; $y > 1$.

Number patterns and sequences

> ⟪ A sequence is a pattern of numbers that grows according to a mathematical rule.
>
> ⟪ You can generate the terms of a sequence using a term-to-term rule or a position-to-term rule.

A Standard number patterns

1 Each number in a sequence is called a **term**.

2 There are several patterns of numbers you need to be able to recognise when you see them. The '…' symbol means that the sequence carries on forever.

Name	Terms	Dot pattern
Even numbers	2, 4, 6, 8, 10, …	
Odd numbers	3, 5, 7, 9, 11, …	
Multiples of 3	3, 6, 9, 12, 15, …	
Square numbers	1, 4, 9, 16, 25, …	
Cube numbers	1, 8, 27, 64, 125, …	
Triangular numbers	1, 3, 6, 10, 15, …	
Powers of 2	2, 4, 8, 16, 32, 64, …	
Powers of 10	10, 100, 1000, 10 000, 100 000, 1 000 000, …	
Fibonacci numbers	0, 1, 1, 2, 3, 5, 8, 13, 21, …	

B Sequences from diagrams

1 You might be asked to produce a sequence of numbers from patterns in a diagram. You might also be asked to draw the next diagram in the sequence or predict the next number.

2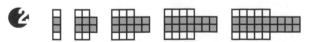

These diagrams generate the following sequences:

White squares: 2, 4, 6, 8, 10, …

Pink squares: 2, 6, 10, 14, 18, …

Totals: 4, 10, 16, 22, 28, …

C Term-to-term rules

1 A **term-to-term** rule tells you how to find the next number in a sequence.

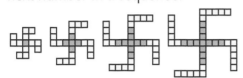

These diagrams generate the following sequences:

Green squares: 1, 5, 9, 13, …
Yellow squares: 12, 16, 20, 24, …

In both cases, the term-to-term rule is 'add 4'. The sequences are different because their first terms are different.

2 For the sequence of powers of 2, the term-to-term rule is 'multiply by 2'. However, starting with a number other than 2 produces a new sequence: 3, 6, 12, 24, 48, …

3 More complex rules are possible. The rule 'divide by 2, then add 1', starting at 10, produces a sequence of decimals: 10, 6, 4, 3, 2.5, 2.25, 2.125, …

4 Fibonacci sequences use the rule 'add the last two terms to find the next term'. You can start with any two numbers (e.g. 4, 7, 11, 18, 29, …), but the most common one is given in section A.

D Position-to-term rules

1 You can generate the terms of a sequence from their **positions** in the sequence. The letter n is normally used for the position, and u_n for the nth term or general term.

>> **key fact** The small number to the right of u is a suffix. It is just a sort of label and has nothing to do with indices.

Position	Calculation	Term
1	$2 \times 1 - 1$	$u_1 = 1$
2	$2 \times 2 - 1$	$u_2 = 3$
3	$2 \times 3 - 1$	$u_3 = 5$
n	$2 \times n - 1$	$u_n = 2n - 1$

The final cell in the table shows the position-to-term rule written in algebra. By substituting different numbers for n, you can find the value of any term in the sequence.
For example, for the 100th term, substitute $n = 100$: $u_{100} = 2 \times 100 - 1 = 199$.

2 The position-to-term rules for some of the sequences in section A are given below.

Even numbers	$u_n = 2n$
Multiples of 3	$u_n = 3n$
Square numbers	$u_n = n^2$
Cube numbers	$u_n = n^3$
Triangular numbers	$u_n = \dfrac{n(n+1)}{2}$
Powers of 2	$u_n = 2^n$
Powers of 10	$u_n = 10^n$

Check by substituting that they give the right results.

>> practice questions

1 Find the missing term(s) in each sequence.
(a) 4, 7, 10, ■, 16, …   (b) 10, 6, 2, ■, …
(c) 5, 6, 8, 11, ■, …   (d) 83, ■, 74, 65, 53, …
(e) ■, ■, 20, 40, 80, …   (f) 5, 9, 14, 23, ■, …

3 Write down the first four terms of these sequences.
(a) $u_n = 4n - 2$ (b) $u_n = n^2 + 10$
(c) $u_n = 3n$ (d) $u_n = \dfrac{n(n+1)(n+2)}{3}$

2 Write down the first five terms of these sequences.

	(a)	(b)	(c)	(d)
First term	4	5	50	3
Term-to-term rule	add 10	subtract 4	divide by 2	multiply by 3 then subtract 4

Sequences and formulae

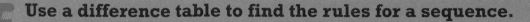

🔧 **Use a difference table to find the rules for a sequence.**

🔧 **Some sequences are created from standard number patterns by a simple operation.**

🔧 **To find the formula for a sequence of fractions, analyse the numerators and denominators separately.**

A Difference tables

1 A **difference table** can make it easier to see how the numbers in a sequence are produced.

>> key fact A difference table should show the positions, terms and differences.

positions, $n =$ ① ② ③ ④ ⑤ ...

terms, $u_n =$ 2 5 8 11 14 ...

difference row +3 +3 +3 +3

This shows that the terms increase by 3 each time.

B Linear sequences

1 **key fact** If the numbers in the difference row are all the same, you are dealing with a linear sequence. This means that if you plotted the terms against their positions on a graph, all the points would be on a straight line.

With a linear sequence, you can write down the term-to-term rule immediately. The rule for the sequence in section A is 'first term 2, add 3'.

2 The position-to-term rule must contain $3n$ in the formula to make all the differences 3. However, although this makes the terms increase in the right way, it doesn't give the first term as 2.

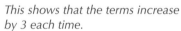

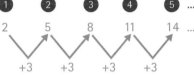

$n =$ ① ② ③ ④ ⑤ ...

$u_n =$ 22 14 6 −2 −10 ...

−8 −8 −8 −8

This shows that the terms decrease by 8 each time.

Here the differences are negative because the terms decrease. The term-to-term rule for this sequence is 'first term 22, subtract 8'. The position-to-term rule needs to have $-8n$ in it. The formula for this sequence is
$u_n = -8n + 30$

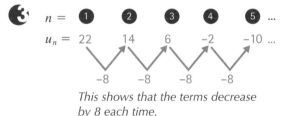

positions, $n =$ ① ② ③ ④ ⑤ ...

$3n =$ 3 6 9 12 15 ...

terms, $u_n =$ 2 5 8 11 14 ... −1

The numbers are all 1 less than a multiple of 3, so the formula is $u_n = 3n - 1$.

C Other sequence types

1 Sometimes, you may need a second difference row to see what is happening.

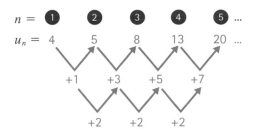

You could describe this as 'first term 4, add 1, 3, 5, 7, etc.' Sequences like this do have position-to-term formulae, but you are not required to find them in the Higher tier exam.

If the differences don't reveal a simple pattern, check these two possibilities.

2

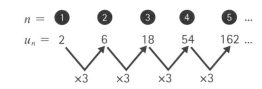

This is a sequence involving powers, as you multiply by the same number each time. The term-to-term rule is 'first term 2, multiply by 3'. The position-to-term formula is

$$u_n = \frac{2 \times 3^n}{3}, \text{ or } u_n = 2 \times 3^{(n-1)}.$$

3

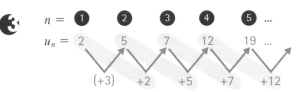

If the differences turn out to be the same as the terms, you are dealing with a Fibonacci-type sequence. The rule is 'add the previous two terms', and you have to state the first **two** terms. This can also be written $u_n = u_{n-1} + u_{n-2}$.

D Variations on standard patterns

1 Some sequences can be built up by altering the standard number patterns slightly.

Examples:

2, 5, 10, 17, 26, …	square numbers with 1 added	$u_n = n^2 + 1$
2, 8, 18, 32, 50, …	square numbers doubled	$u_n = 2n^2$
11, 18, 37, 74, 135, …	cube numbers with 10 added	$u_n = n^3 + 10$
20, 40, 80, 160, 320, …	powers of 2 multiplied by 10	$u_n = 10 \times 2^n$

E Sequences of fractions

1 Analyse the numerator and denominator separately, then combine them into a single formula.

Example:
$$\frac{2}{1}, \frac{5}{4}, \frac{8}{9}, \frac{11}{16}, \frac{14}{25}, \dots$$

The numerators are the terms of the sequence in section A, $u_n = 3n - 1$.

The denominators are square numbers, $u_n = n^2$.

So the fraction sequence is $u_n = \frac{3n-1}{n^2}$.

>> practice questions

Each question gives the first five terms of a sequence. For each one:

(a) Draw a difference table and find a term-to-term rule (if possible).

(b) Find a position-to-term formula (if possible).

(c) Calculate the 10th term.

1 6, 11, 16, 21, 26, …

2 3, 9, 15, 21, 27, …

3 10, 9, 8, 7, 6, …

4 1, 1.2, 1.4, 1.6, 1.8, …

5 2, 5, 9, 14, 20, …

6 12, 48, 192, 768, 3072, …

7 0, 3, 8, 15, 24, …

8 2, 16, 54, 128, 250, …

9 11, 17, 28, 45, 73, …

10 $\frac{1}{2}, \frac{3}{4}, \frac{5}{8}, \frac{7}{16}, \frac{9}{32}, \dots$

Lines and equations

 Every straight line on a co-ordinate grid has an equation satisfied by all the points on it.

 The gradient of a line is its slope or steepness, and is equal to the coefficient of x in the equation.

 All lines with the same gradient are parallel. The position of a particular line depends on the y-intercept, the value of y when $x = 0$

A Graphs of straight lines

Every straight line on a co-ordinate grid can be written in the form $y = mx + c$.

So in the equation $y = 2x - 3$, $m = 2$ and $c = -3$.

>> **key fact** m **is the number multiplying** x **(the coefficient of** x**), and represents the gradient or steepness of the line.** c **gives the position where the line crosses the** y**-axis and is called the** y**-intercept of the line.**

B Calculating gradient

1 To calculate the gradient of a line on a diagram:

- select two points on it;

- work out the increase in x from one point to the other, and the increase in y;

- $\dfrac{\text{increase in } y}{\text{increase in } x}$ gives the gradient.

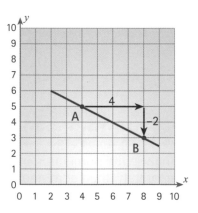

2 Find the gradient of the line in the diagram.

The two chosen points are $A(4, 5)$ and $B(8, 3)$.

From A to B, the increase in x is 4 units, and the increase in y is -2 units.

The gradient is $m = \dfrac{\text{increase in } y}{\text{increase in } x} = \dfrac{-2}{4} = -\dfrac{1}{2}$

C Finding the equation of a line

1 Once the gradient of a line is known, its equation can be found by substituting x and y from a known point on the line. This will determine the value of c.

2 Find the equation of the line in the diagram above, and determine where it crosses the y-axis.

You know that $m = -\dfrac{1}{2}$. So the equation of the line is $y = -\dfrac{1}{2}x + c$.

Pick one of the points on the line, say (4, 5). Substitute for x and y:

$y = -\frac{1}{2}x + c$, so $5 = -\frac{1}{2} \times 4 + c$

$5 = -2 + c$, so $c = 7$.

So $y = -\frac{1}{2}x + 7$. This means the line crosses the y-axis at (0, 7).

D Equations of the form $ax + by = k$

1 Equations of this type also represent straight lines.
Examples are $3x + y = -2$, $2x - 5y = 0$, etc.
To find the y-intercept, substitute $x = 0$ into the equation.

Find the y-intercept of the line with equation $3x - 2y = 2$.

Substitute $x = 0$: $3x - 2y = 2$

$$3 \times 0 - 2y = 2$$

$-2y = 2$, so $y = -1$.

The line crosses the y-axis at (0, −1).

2 To determine the gradient, make y the subject of the equation and use the coefficient of x.

Find the gradient of the line with equation $3x - 2y = 2$.

First, make y the subject: $y = \frac{3x - 2}{2} = \frac{3}{2}x - 1$.

The gradient of the line is $\frac{3}{2}$ (or 1.5).

Note that the value of c can also be found from the rearranged equation.

E Perpendicular lines

1 Two lines that intersect at right angles have the following property: the product of their gradients must equal −1.

2 The equation of the line perpendicular to $y = 3x - 5$,
passing through the point (2, 1), is $y = -\frac{1}{3}x + \frac{5}{3}$.

>> practice questions

1 **Write the gradient and y-intercept of lines with the equations:**

   (a) $y = 7x + 3$ (b) $y = 3x - 5$ (c) $y = x + 12$ (d) $y = 3x - 2$

2 **Find the gradient and y-intercept of the lines with the following equations:**

   (a) $3x + y = 4$ (b) $2x - y = 7$ (c) $x - 3y = 7$ (d) $x + 5y = 9$

3 **The gradient of a line is 4 and its y-intercept is at (0, 5). What is the equation of the line?**

4 **A line passes through (0, 6) and (2, 8). Find the equation of the line.**

5 **Find the equation of the line that passes through the point (0, 1) and is parallel to $y = 2x - 1$.**

6 **Find the equation of the line that is perpendicular to $y = 8 - 2x$, and intersects it at the point where $x = 3$.**

Curved graphs

- The shape of a quadratic graph is a curve called a parabola. The coefficient of x^2 determines which way up the parabola is, and how steep it is.

- You need to be familiar with the shapes of quadratic, cubic and reciprocal graphs.

A The graph of $y = x^2$

key fact The shape of the graph of $y = x^2$ is called a parabola.

Notice that it has the y-axis as a line of symmetry, and passes through the origin.

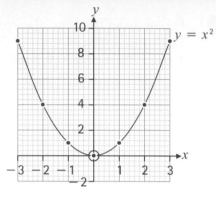

The graphs of $y = 2x^2$, $3x^2$ etc. look similar, but are steeper. If negative coefficient of x^2 is involved, the graph turns upside down.

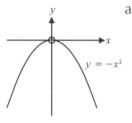

B Other quadratic graphs

The equation of a quadratic graph is generally of the form $y = ax^2 + bx + c$. These graphs don't have to pass through the origin and may have a different line of symmetry, but they are all parabolic. As with linear graphs, c represents the y-intercept.

Example: plot the graph of $y = x^2 - 4x + 3$, for values of x from -4 to 4.

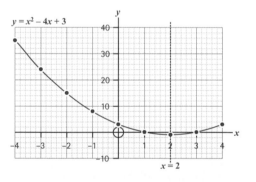

The line of symmetry of this graph is $x = 2$.

x	−4	−3	−2	−1	0	1	2	3	4
$y = x^2 - 4x + 3$	35	24	15	8	3	0	−1	0	3

C Cubic graphs

Cubic graphs have equations of the form
$y = ax^3 + bx^2 + cx + d$. If a is positive, the graph has
one of these two shapes:

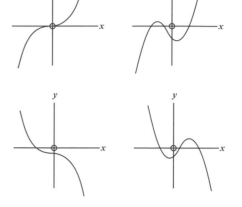

If a is negative, the graph has one of these two shapes:

d represents the y-intercept.

D Reciprocal graphs

Reciprocal graphs have equations of the form
$y = \frac{a}{x}$. The graph has one of these two shapes,
depending on the sign of a:

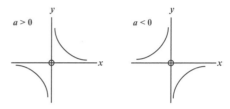

E Solving equations graphically

1 An equation of the form f$(x) = 0$ can be solved by plotting the graph
of $y = $ f(x) and finding the points where it crosses the x-axis.

The graph of $y = x^2 + 3x - 3$ is the red curve on this grid.

The solutions to the equation $x^2 + 3x - 3 = 0$ are found where the
graph crosses the x-axis. They are $x = -2.7$, -0.7 and 2, to 1 dp.

2 To solve $x^2 + 3x - 13 = 0$, add 10 to both sides: $x^2 + 3x - 3 = 10$.
So the intersections of $y = x^2 + 3x - 3$ with $y = 10$ (the pink line on
the grid) solve this equation. There is only one, when $x = 2.8$ (1 dp).

To solve $x^2 + 2x - 4 = 0$, add $x + 1$ to both sides:
$x^2 + 3x - 3 = x + 1$. So the intersections of $y = x^2 + 3x - 3$ with
$y = x + 1$ (the green line on the grid) solve this equation. There are
three, when $x = -2.8$, 0.4 and 2.4 (1 dp).

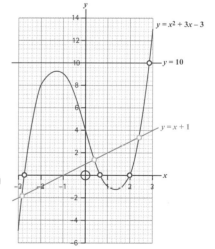

>> practice questions

1 **Solve these equations graphically, giving your answers correct to 1 dp.**

  (a) $x^2 - 2x - 1 = 0$ (x-axis: -3 to 5; y-axis: -5 to 15)

  (b) $3x^2 - 10 = 0$ (x-axis: -3 to 3; y-axis: -10 to 20)

2 **Add extra lines to the graphs from question 2 to solve these equations.**

  (a) $x^2 - 2x - 6 = 0$

  (b) $3x^2 + 2x - 15 = 0$

Transforming graphs

Vertical and horizontal translations, stretches and squashes of graphs have simple algebraic representations.

A Vertical translations

Suppose you have a graph of a function, $y = f(x)$. The graph can be translated up or down by adding or subtracting a number.

>> **key fact** To translate $y = f(x)$ by a units in the positive y direction, replace it by $y = f(x) + a$. To translate it by a units in the negative y direction, replace it by $y = f(x) - a$.

Here is the graph of $y = x^2$, translated 3 units in the negative y direction to become $y = x^2 - 3$:

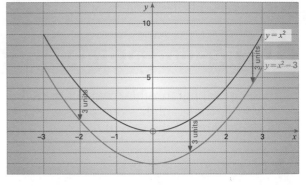

B Horizontal translations

To translate the graph of $y = f(x)$ by 1 unit in the positive x direction, replace x by $x - 1$.

>> **key fact** To translate $y = f(x)$ by a units in the positive x direction, replace it by $y = f(x - a)$. To translate it by a units in the negative x direction, replace it by $y = f(x + a)$.

So the graph of $y = \frac{1}{x-3}$ is the same as the graph of $y = \frac{1}{x}$, translated 3 units in the positive x direction:

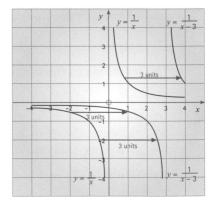

C Vertical stretches

The graph of $y = f(x)$ can be stretched vertically by multiplying every y value by some number a. If $a < 1$, the graph is 'squashed'.

>> **key fact** In general, to stretch $y = f(x)$ vertically by a factor of a, replace it by $y = af(x)$.

So the graph of $y = 3\cos x$ is the same as the graph of $y = \cos x$, stretched vertically 3 times. Note that points of the graph that lie on the x-axis are unaffected by the stretch.

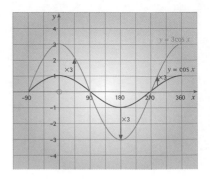

D Horizontal squashes

1 To **squash** the graph of $y = f(x)$ by a factor of 3 horizontally, replace x by $3x$.

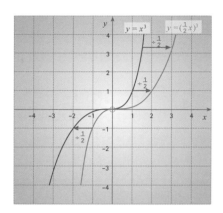

>> **key fact** In general, the graph of $y = f(ax)$ is the same as the graph of $y = f(x)$, horizontally squashed by a factor of a. If $a < 1$, the transformation is a stretch and increases the x scale.

2 To **stretch** the graph of $y = f(x)$ by a factor of 2 horizontally, replace x by $\frac{1}{2}x$.

Here is the graph of $y = x^3$, stretched by a factor of 2 horizontally to become $y = (\frac{1}{2}x)^3 = \frac{x^3}{8}$:

Note that points of the graph that lie on the y-axis are unaffected by the stretch.

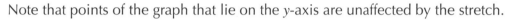

practice questions

1 Write down the equation of the graph formed when $y = x^2$ is translated:

(a) 4 units up

(b) 3 units left

(c) 5 units down

(d) 7 units right.

2 For each part, write down the equation of the image graph.

	object graph	transformation
(a)	$y = x^3$	$3 \times$ vertical stretch
(b)	$y = x^2$	squash to half size vertically
(c)	$y = x^2$	$2 \times$ horizontal stretch
(d)	$y = (x + 1)^2$	squash to $\frac{1}{10}$ size horizontally
(e)	$y = x^3 + 5x^2$	$5 \times$ horizontal stretch

3 For each part, describe the transformation given by the change in equation.

	object graph	image graph
(a)	$y = x^2$	$y = 5x^2$
(b)	$y = x^2$	$y = (\frac{x}{3})^2$
(c)	$y = x^2 + 2x$	$y = 4x^2 + 4x$
(d)	$y = \tan x$	$y = \tan(\frac{x}{2})$
(e)	$y = \frac{1}{x}$	$y = \frac{10}{x}$

Linear simultaneous equations

- A pair of equations containing two unknown letters, that are both true, are called simultaneous equations.

- By combining equations, it's possible to make one of the unknown letters disappear, leaving a simple equation with just one letter to find. This is called elimination.

- If the equations are linear, their graphs are straight lines. The coordinates of the point of intersection give the solution to the simultaneous equations.

A Elimination

Sometimes it's possible just to add two equations together and perform an **elimination**.

Example:

$$x - y = 3 \qquad ①$$
$$x + y = 11 \qquad ②$$

Note that the two equations have each been given a number, to make them easy to refer to.

So $x - y + x + y = 3 + 11$ *add the two equations*

$$2x = 14 \qquad \textit{simplify: y has been eliminated}$$

$$x = 7 \qquad \textit{divide by 2}$$

Substitute x into ①, to find the value of y.

$$7 - y = 3$$
$$y = 4$$

Check: substitute x and y into equation ②:

$$7 + 4 = 11 ✓$$

>> **key fact** Always use one equation for the calculation and the other equation for the check.

B Multiplying equations

1 You can only eliminate by adding or subtracting equations if the coefficients match. Sometimes, you need to 'force' a match by multiplying one or both of the equations first.

2 *Example:* Solve $3x + 2y = 46$ ①

$$2x - 5y = 18 \quad ②$$

To eliminate x, multiply equation ① by 2 and equation ② by 3:

$$6x + 4y = 92 \qquad ③ \qquad ① × 2$$
$$6x - 15y = 54 \qquad ④ \qquad ② × 3$$

Now subtract equation ④ from equation ③:

$$19y = 38 \qquad ③ - ④: \text{eliminates } x$$

so $y = 2$ *divide by 2*

Substitute y into equation ①:

$$3x + 4 = 46 \qquad \textit{substitute}$$
$$3x = 42 \qquad \textit{subtract 4}$$
$$x = 14 \qquad \textit{divide by 3}$$

Check: substitute for x and y in equation ②:
$28 - 10 = 18 ✓$

3 Notice that you could also solve these equations another way: multiply equation ① by 5 and equation ② by 2, then add the resulting equations together to eliminate y.

C Using a graph

1. The graphs of linear equations are straight lines. To solve a pair of simultaneous equations, plot their graphs on the same axes.

>> **key fact** The solution is given by the co-ordinates of the point where the lines intersect.

2. *Example*: Solve the simultaneous equations $x + y = 4$ and $6y + 2x = 12$.

First, make y the subject of both equations.

$x + y = 4$

$y = 4 - x$ *subtract x from both sides*

This graph can now be plotted.

$6y + 2x = 12$

$6y = 12 - 2x$ *subtract 2x from both sides*

$y = 2 - \frac{1}{3}x$ *divide by 6*

Now plot this graph.

The graph shows that the lines cross at (3, 1), so the solution to the equations is $x = 3$ and $y = 1$.

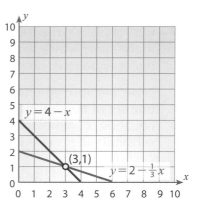

>> practice questions

1. **Solve each pair of simultaneous equations using an algebraic method.**

(a) $x + y = 7$ $x - y = 3$

(b) $3m + n = 19$ $m + n = 7$

(c) $3t + r = 24$ $t - 2r = -6$

(d) $3x + 2y = 23$ $2x + y = 14$

(e) $4c + 2d = 46$ $3c + 3d = 48$

2. **Solve each pair of simultaneous equations by drawing a graph.**

(a) $x + y = 10$ $2x + y = 13$

(b) $x + 3y = -1$ $3x - y = 9$

Mixed simultaneous equations

- Simultaneous equations in the exam will be one of two types: both linear or one linear and one quadratic.

- When one of the equations is quadratic, there will usually be two sets of solutions.

A Solving equations 1

1 Solve the equations:

$$x^2 + 2y^2 = 30$$
$$x + y = 5.$$

2 First of all, rewrite them and label them in the same way you do with simultaneous linear equations. So you have:

$$x^2 + 2y^2 = 30 \quad ①$$
$$x + y = 5 \quad ②$$

3 Now, because you have **two unknowns and one set of quadratics**, you need to rearrange equation ②, either in terms of x or y and then substitute that value into equation ①.

4 Here you are going to rearrange equation ② to make y the subject, because this is the easier substitution. From equation ②:

$$y = 5 - x \quad ③$$

5 So now substitute for y into equation ①. This now becomes:

$$x^2 + 2(5 - x)^2 = 30$$
$$x^2 + 2[(5 - x)(5 - x)] = 30$$
$$x^2 + 2[25 - 10x + x^2] = 30$$
$$x^2 + 50 - 20x + 2x^2 = 30$$
$$3x^2 - 20x + 50 = 30$$
$$3x^2 - 20x + 20 = 0$$

6 Now you can solve this using the quadratic formula.

>> **key fact** **You need to memorise the formula and be able to quote it.**

$$x = \frac{-b \pm \sqrt{b^2 - 4ac}}{2a}, \text{ where } a = 3, b = -20, c = 20$$

7 $x = \frac{20 \pm \sqrt{160}}{6}$, so $x = 1.23$ or $x = 5.44$, to 2 dp.

8 When $x = 1.23$, $y = 3.77$, when $x = 5.44$, $y = -0.44$.

B Solving equations 2

❶ Solve the equations:

$$y = 5x^2 \qquad ①$$

$$x - y = 0 \qquad ②$$

❷ Rearrange equation ② to make x the subject. $x = y$, so substituting this in equation ①, $y = 5y^2$.

❸ Rearrange: $5y^2 - y = 0$.

❹ Factorise: $y(5y - 1) = 0$.

❺ So either $y = 0$ or $5y - 1 = 0$, $y = \frac{1}{5}$.

❻ When $y = 0$, $x = 0$. When $y = \frac{1}{5}$, $x = \frac{1}{5}$.

C Repeated roots

❶ Quadratic equations may have repeated roots, so some pairs of simultaneous equations have them, too.

❷ Consider the equations

$$x^2 + y^2 = 25 \qquad ①$$

$$3x = 25 - 4y \qquad ②$$

Make x the subject of ②: $\qquad x = \frac{25 - 4y}{3}$

❸ Substitute this into ①:

$$\left(\frac{25 - 4y}{3}\right)^2 + y^2 = 25$$

So:

$$\frac{(25 - 4y)^2}{9} + y^2 = 25$$

Multiply by 9:

$$(25 - 4y)^2 + 9y^2 = 225$$

Expand:

$$625 - 200y + 16y^2 + 9y^2 = 225$$

So:

$$25y^2 - 200y + 625 = 225$$

and

$$25y^2 - 200y + 400 = 0$$

Divide by 25:

$$y^2 - 8y + 16 = 0$$

Factorise:

$$(y - 4)^2 = 0$$

So there is a repeated root, $y = 4$.

❹ If $y = 4$, then $3x = 25 - 16 = 9$, so $x = 3$.

The solution is $x = 3$, $y = 4$, repeated.

>> practice questions

Solve the following, if possible:

1 $x^2 + y^2 = 25$
   $x + y = 7$

2 $xy = 5$
   $2x - y = -3$

3 $y^2 = 2x$
   $x - y = 0$

4 $x^2 - y^2 = 20$
   $x + y = 7$

5 $2x^2 + 5y = 38$
   $x + y = 7$

Rounding and accuracy

- Rounded numbers are used when limited accuracy is required.
- To estimate the result of a calculation, carry it out using rounded numbers.
- A measurement accurate to the nearest unit may be out by up to half a unit above or below.

A Rounded numbers

1 *Rounding to the nearest 10, 100, etc.*

The number 24 is closer to 20 than it is to 30, so 24 rounded to the nearest ten is 20. 25 is exactly halfway. In a 'halfway' situation, you always round **up**. So 25 rounded to the nearest ten is 30.

2 *Rounding to a given number of decimal places (dp)*

Suppose you needed to round 1.428 571 4 to 2 dp.

- Split the number after the required number of decimal places:

>> **key fact** If the 'next' digit is 5 or over, round up.

1.42 | 857 14

- Look at the first digit after the split.

  It's 8. This is over 5, so you round the number **up** to 1.43.

3 *Rounding to a given number of significant figures (sf)*

Here, the type of rounding depends on the size of the number. Count digits from the first non-zero digit in the number, and split after this. The table shows some different cases.

Number	Round to	Split number	Type of rounding	Rounded number
6415	2 sf	64|15	nearest hundred	6400
0.06666666	3 sf	0.0666|6666	4 dp	0.0667
84.9	1 sf	8|4.9	nearest ten	80

B Estimating answers

1 Sometimes it is useful to check a calculation by making an **estimate** of the answer.

>> **key fact** Round the numbers in the calculation to 1 sf first.

2 To estimate the answer to $\frac{235 \times 7.81}{38.33}$, round the numbers so the calculation becomes $\frac{200 \times 8}{40}$. Then calculate with these numbers:

$\frac{200 \times 8}{40} = \frac{1600}{40} = 40$.

3 The actual answer using the original numbers is $\frac{1835}{38.33} = 47.88$ to 2 dp.

This is close to the estimate, so you can be confident it's correct.

C Accuracy of measurements

1 Suppose that you are told that a length is 63 mm, correct to the nearest millimetre. There is a range of actual measurements that could have produced this figure. Anything 63.5 mm or above would round up to 64 mm. Anything below 62.5 mm would round down to 62 mm. So the range is 62.5 mm ⩽ length < 63.5 mm. Notice that 62.5 mm is allowed, but 63.5 is not.

2 Here are some further examples showing different levels of accuracy. The symbol ± means 'plus or minus'.

length	accuracy	tolerance	range allowed
450 mm	nearest 10 mm	± 5 mm	445 mm ⩽ length < 455 mm
98 m	nearest metre	± 0.5 m	97.5 m ⩽ length < 98.5 m
6.33 km	2 decimal places	± 0.005 km	6.325 km ⩽ length < 6.335 km
210 cm	nearest 5 cm	± 2.5 cm	207.5 cm ⩽ length < 212.5 cm

>> **key fact**  The two limits for a measurement are called the **upper bound** and **lower bound**.

>> **key fact** If a measurement is given to the nearest unit, the tolerance is half a unit each way.

D Calculating with rounded measurements

1 **key fact** If you use rounded measurements in a calculation, a range of answers is possible. Suppose you had a rectangle 8 cm by 6 cm, but the measurements were known to be accurate to only the nearest centimetre. What range of area could the rectangle have?

7.5 cm ⩽ length < 8.5 cm
5.5 cm ⩽ length < 6.5 cm

The minimum possible area is
$7.5 \times 5.5 = 41.25 \text{ cm}^2$.

The maximum possible area is
$8.5 \times 6.5 = 55.25 \text{ cm}^2$.

So $41.25 \text{ cm}^2 ⩽ \text{area} < 55.25 \text{ cm}^2$.

2 If you need to subtract or divide, be careful in choosing the measurements to use: suppose you need to divide A by B.

Upper bound of answer =
upper bound (A) ÷ lower bound (B)

Lower bound of answer =
lower bound (A) ÷ upper bound (B)

The same rule works with subtraction.

>> practice questions

1 Estimate the answer to each calculation, then find the exact answer, rounded to 3 sf.

(a) $(31.42 - 15.7) \times 2.25$

(b) $\frac{13.7 \times 35.1}{14}$ (c) $\frac{3.57}{1.81} \times \frac{2.26}{4.009}$

2 Write down the range of measurements for each rectangle, then calculate its smallest and largest possible area.

(a) 10 cm by 5 cm, accurate to the nearest cm

(b) 16 mm by 6.5 mm, accurate to the nearest 0.5 mm

3 The volume of water in a tank is known to be 1000 litres, correct to the nearest 10 litres. The tank is a cylinder, and the cross section is 0.8 m², correct 1 decimal place.

Find the upper and lower bounds for the:

(a) volume of water,

(b) area of cross section of tank,

(c) depth of water in tank.

Dimensions

A Length

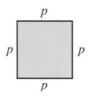 **key fact** You can check if a formula is describing length by looking at its components.

♩ Consider a square of side length p. The formula for the perimeter is $4p$.

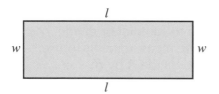

This formula obviously represents a length, because p is a length and 4 is just a number.

♬ A rectangle's perimeter will also have a formula for the sum of the lengths of its sides.

The formula here becomes $2l + 2w$ or $2(l + w)$.

Again, this is a number multiplied by a length.

♪ A circle of diameter d units has a perimeter of πd or $2\pi r$.

♪ Rob says that the formula $3\pi r$ represents a length, but John argues that it cannot be so. Who is correct?

3 and π are numbers, they do not have a dimension, therefore r is the only element of the formula that does have a dimension. That means this formula is of **dimension one**, so it *may* be a length.

B Area

>> **key fact** **Whenever you have a formula for area, you will always have: length × length.**

♪ The formula for the area of a circle is πr^2.

In the analysis you can ignore π because it is a number and therefore has no dimension, but you do have $r \times r$ in this formula.

♪ In other words, you have an area. So the formula is of **dimension two**.

C Volume

>> **key fact** **All formulae for volume have the dimensions length × length × length.**

♪ The formula for the volume of a cylinder is $\pi r^2 h$.

Again we can simply ignore the numbers, in this case π, so the dimensions to consider are r^2 and h.

So this is written as $r \times r \times h$, so it must be a volume.

♪ *Example*:

$5\pi r^2$ $\frac{4}{3}\pi r^3$ $\frac{3}{5}\pi r^2$.

Which one is a volume?

First of all, ignore the numbers because they do not have dimensions.

So in $5\pi r^2$ we need to consider $r \times r$. This is **dimension two**, so it must be an area.

This is also true for $\frac{3}{5}\pi r^2$.

In $\frac{4}{3}\pi r^3$, we need to consider $r \times r \times r$. This is **dimension three**, so this must be a volume.

>> practice questions

1 **The letters h and r represent lengths. For each of the following formulae, write down if it represents a length, an area or a volume.**

   (a) $2\pi r$ **(b) πr^2**

   (c) $\pi r^2 h$ **(d) $2\pi rh$**

2 **The letters l and w represent lengths. Explain why l^2lw cannot represent an area.**

3 **Explain what the formula $\frac{4}{3}\pi r^3$ represents.**

Speed and motion graphs

> **C** Average speed = $\frac{distance}{time}$; use the DST triangle to remember the relationships between distance, speed and time.

A Speed

1 During a journey, the speed of a moving object changes as it accelerates and decelerates. Exam questions are usually about the **average** speed for a journey.

>> **key fact**

Average speed = $\frac{distance}{time}$

Example:

A train takes 2 hours to cover a distance of 130 km. What is its average speed?

Speed = $\frac{130}{2}$ = 65 km/h.

2 When working with speed, you have to be careful that the units match.

Example:

Ellie cycles 3 km to school in 15 minutes. What is her average speed in km/h?

15 minutes = $\frac{1}{4}$ hour (match the time units in the question and the answer)

Speed = 3 ÷ $\frac{1}{4}$ = 3 × 4 = 12 km/h.

You may need to convert one unit of speed to another. For example, 1 metre per second = 60 metres per minute = 3600 metres per hour = 3.6 km/h.

B Time and distance

1 There are two other formulae you may need to use:

>> **key fact**

time = $\frac{distance}{speed}$

distance = speed × time

2 You can remember this using the DST triangle:

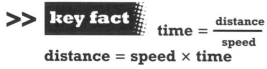

To use it, cover up the letter for the quantity you're trying to find. The other two are in the right relationship, i.e. this one shows

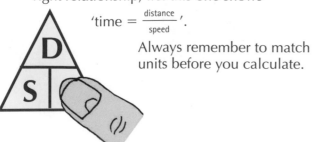

'time = $\frac{distance}{speed}$'.

Always remember to match units before you calculate.

Example:

Adil drives for $3\frac{1}{2}$ hours at an average speed of 50 mph. How far was his journey?

Distance = 50 × 3.5 = 175 miles.

Example:

The average speed of Harry's model train is 0.7 m/s. The track is 35 m long.

How long does it take the train to complete a circuit of the track?

Time = 35 ÷ 0.7 = 50 seconds.

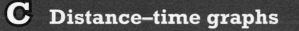

C Distance–time graphs

The graph illustrates a bus journey.

Distances east of Rhyl are shown as positive, those to the west as negative. The bus travels the 5 miles from Rhyl to Prestatyn in 10 minutes.
The average speed of the bus is 5 miles ÷ $\frac{1}{6}$ hour = 30 mph.

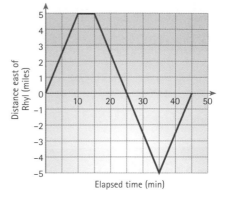

>> **key fact** On a distance–time graph, the gradient represents the speed. Horizontal lines show stationary parts of the journey.

The portion of the graph that has a negative gradient still shows a speed of 30 mph, but in the opposite direction. You can tell from the graph that the bus waited for 10 minutes in Prestatyn, but turned round immediately in Abergele for the return journey.

D Speed–time graphs

① This graph shows the journey of a tube train between two stations.

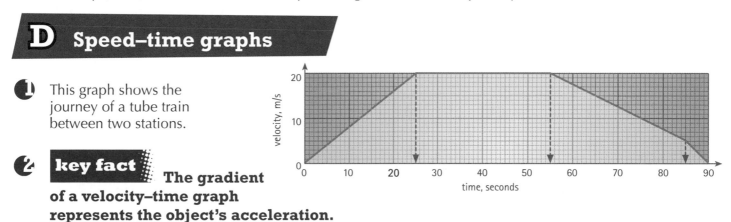

② **key fact** The gradient of a velocity–time graph represents the object's acceleration.

For example, between 55 and 85 seconds, the gradient is:

$\frac{\text{increase in velocity}}{\text{increase in time}} = \frac{-15\,\text{m/s}}{30\,\text{sec}} = -0.5$ m/s² (negative because the train is decelerating).

③ **key fact** The area enclosed by a velocity–time graph and the time-axis gives the distance travelled.

Area $= \frac{1}{2}(25 \times 20) + 30 \times 20 + \frac{1}{2}(30 \times (20 + 5)) + \frac{1}{2}(5 \times 5) = 1237.5$ m

>> practice questions

1 **Calculate the missing value for each of these journeys.**

	distance	time	average speed
(a)	30 km	2 hours	(km/h)
(b)	4 km	30 minutes	(km/h)
(c)	100 km	(hours)	40 km/h
(d)	(metres)	0.05 seconds	250 m/s
(e)	(km)	1 hour	8 km/second
(f)	400 m	1 minute	(km/h)

2 **A car accelerates from rest at 5m/s² for 15 seconds, then moves with a constant speed for 15 seconds, and finally decelerates at 3m/s² back to rest.**

(a) Draw a velocity-time graph for the journey of the car.

(b) Calculate the total distance travelled.

Properties of shapes

- The exterior angles of any polygon add up to 360°.
- The angle sum of any polygon is a multiple of 180°.
- Regular polygons have all sides and angles equal.

A Triangles

1 The angles inside a triangle at the vertices (corners) are called its **interior** angles.

>> **key fact** The interior angles of a triangle add up to 180°.

You make an **exterior** angle by extending one of the sides. In this situation, $x + y = z$.

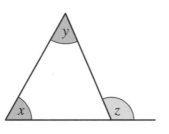

2 **Isosceles** triangles have two sides of equal length and two equal angles. You can use this fact when finding unknown angles.

In this triangle, $2a + 44° = 180°$, so $a = 68°$.

In this triangle, $c + 2 \times 67° = 180°$, so $c = 46°$.

3 All three sides of an **equilateral** triangle are equal and all its angles are 60°.

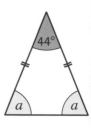
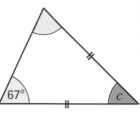

B Quadrilaterals

1 **key fact** The interior angles of a quadrilateral add up to 360°.

There are many different types of quadrilateral. Each type has its own properties.

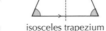

general quadrilateral (no special features) — kite — trapezium — isosceles trapezium — parallelogram — rhombus — rectangle — square

	kite	trapezium	parallelogram	rhombus	rectangle	square
opposite sides	–	1 || pr	2 || pr, 2 = pr	2 || pr, all =	2 || pr, 2 = pr	2 || pr, all =
adjacent sides	2 = pr	–	–	all =	⊥	= and ⊥
opposite angles	1 = pr	–	2 = pr	2 = pr	all = (90°)	all = (90°)
adjacent angles	–	2 supp pr	2 supp pr	2 supp pr	all = (90°)	all = (90°)
diagonals	⊥	–	=	= and ⊥	=	= and ⊥
symmetry	1 line	–	order 2	2 lines, order 2	2 lines, order 2	4 lines, order 4

Key: equal (=), parallel (\|\|), perpendicular (⊥), pair (pr), supplementary (supp).

C Polygons

1 The exterior angles of a polygon always add up to 360°.
One interior angle and its exterior angle always add up to 180°.

This table gives the names of different types of polygon.

Number of sides	5	6	7	8	9	10	12
Name	pentagon	hexagon	heptagon	octagon	nonagon	decagon	dodecagon

2 The angle sum of a polygon is the total of all its **interior** angles. A polygon with n sides has an angle sum of $180(n - 2)°$, or $180n - 360°$. If you split the polygon up into triangles, you can see why.

3 This table gives angle sums for some of the polygons.

Number of sides	3	4	5	6	7	8	10
Angle sum	180°	360°	540°	720°	900°	1080°	1440°

D Regular polygons

1 **key fact** In a regular **polygon, all the sides are the same length, all the interior angles are equal and all the exterior angles are equal.**

In this section, n stands for the number of sides.

2 As all the exterior angles add up to 360°, one exterior angle $= \dfrac{360°}{n}$, and so one interior angle is $180° - \dfrac{360°}{n}$.

3 You can use this to find out how many sides a regular polygon has, if you know one of its interior angles. Suppose one interior angle is 150°. Then the exterior angle $= 180° - 150° = 30°$.

So $\dfrac{360°}{n} = 30°$ and $n = 12$.

The shape is a dodecagon.

This table gives results for some of the polygons.

Number of sides	3	4	5	6	8	10	12
Exterior angle	120°	90°	72°	60°	45°	36°	30°
Interior angle	60°	90°	108°	120°	135°	144°	150°

>> practice questions

1 Find the angle marked x in each diagram.

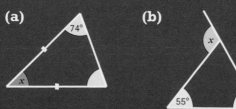

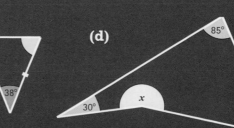

(a) (b) (c) (d)

2 Calculate the angle sum for: **(a) a dodecagon (12 sides)** **(b) an icosagon (20 sides).**

3 The angle sum of a regular polygon is 1260°. Calculate:

(a) the size of one interior angle (b) the size of one exterior angle
(c) the number of sides.

Circle geometry

There are a number of angle facts connected with circles that you need to know. It is possible to deduce most of them from other facts, but this wastes time in an exam.

There are two of these facts you need to be able to prove.

A Tangents and chords

1. The perpendicular bisector of a chord passes through the centre of the circle.

2. The two tangents drawn from a point to a circle are equal.

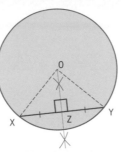

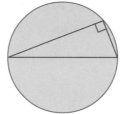

B Angles in semicircles, at the centre, or in the same segment

1. The angle in a semicircle is a right angle.

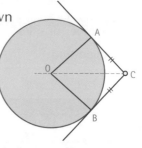

2. The angle subtended by any arc at the centre of a circle is twice the angle subtended by the same arc at any point on the remaining part of the circumference. In this diagram, this means that angle x is double the size of angle y.

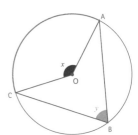

>> **key fact** This is a statement you need to be able to prove.

To do this, draw in the radius OB. This divides quadrilateral OABC into two isosceles triangles. The base angles of $\triangle$OAB have been labelled p, and in $\triangle$OBC, q. So $y = p + q$.

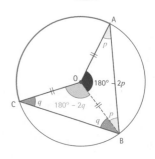

Proof

$\angle AOB$	$= 180° - 2p$	*angles in a triangle total 180°*
$\angle BOC$	$= 180° - 2q$	*angles in a triangle total 180°*
x	$= 360° - (\angle AOB + \angle BOC)$	*angles at a point total 360°*
	$= 360° - (180° - 2p + 180° - 2q)$	
	$= 360° - (360° - 2p - 2q)$	
	$= 2p + 2q$	
	$= 2y.$	

This proves the result.

3. Angles in the same segment are equal.

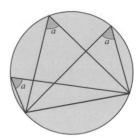

C Cyclic quadrilaterals

Opposite angles in a cyclic quadrilateral are supplementary
(they add up to 180°).

D The alternate segment theorem

1 The angle between a tangent and a chord is equal to the
angle in the alternate segment. In this diagram, this means
that angle x and angle y are equal. The 'alternate segment'
is the part of the circle enclosed by chord PQ, arc PAQ and
containing the centre.

>> **key fact** This is a statement you need to
be able to prove.

2 To do this, draw in the radii OP and OQ.

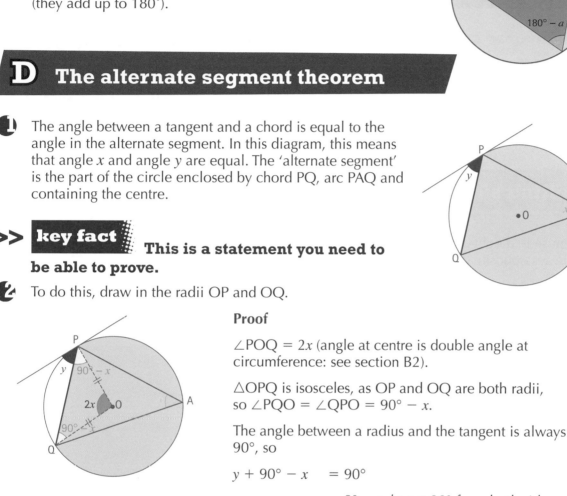

Proof

$\angle POQ = 2x$ (angle at centre is double angle at
circumference: see section B2).

$\triangle OPQ$ is isosceles, as OP and OQ are both radii,
so $\angle PQO = \angle QPO = 90° - x$.

The angle between a radius and the tangent is always
90°, so

$$y + 90° - x = 90°$$

$$y - x = 0° \quad \textit{subtract 90° from both sides}$$

$$y = x. \quad \textit{this proves the result}$$

>> practice questions

1 Find angle a.

2 Find the angles x and y.

3 Find the marked angles.

4 Find angles x, y and z.

Pythagoras' Rule and basic trigonometry

- In a right-angled triangle, the squares of the two shorter sides add up to make the square of the hypotenuse – this is Pythagoras' Rule.

- Use **SOH CAH TOA** to remember the trigonometric ratios.

- Use the inverse trig functions $\sin^{-1}$, $\cos^{-1}$ and $\tan^{-1}$ to find angles.

- Check that your calculator's angle mode is set to degrees, not radians or grades.

A Labelling sides

The **hypotenuse** in a right-angled triangle is its longest side. It is always opposite the right angle.

Once you have chosen an angle to work on (marked x), the other sides are labelled **adjacent** (next to the angle) and **opposite** (not in contact with it).

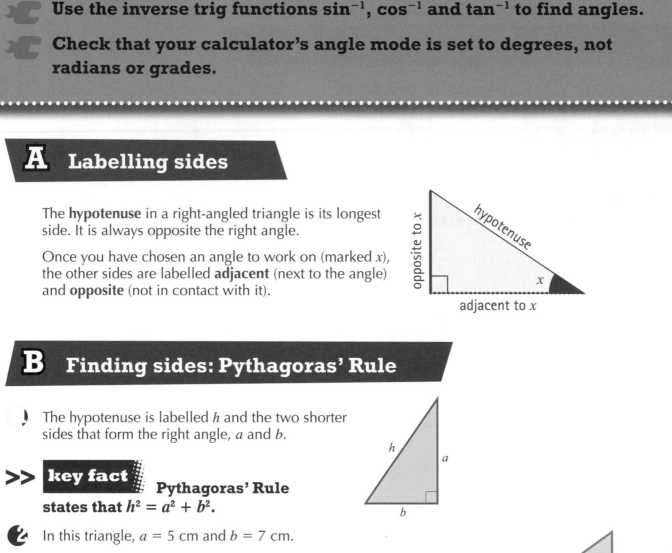

B Finding sides: Pythagoras' Rule

! The hypotenuse is labelled h and the two shorter sides that form the right angle, a and b.

>> **key fact** **Pythagoras' Rule** states that $h^2 = a^2 + b^2$.

In this triangle, $a = 5$ cm and $b = 7$ cm.

By Pythagoras' Rule, $h^2 = a^2 + b^2$

$\qquad = 5^2 + 7^2$ *substitute given values*

$\qquad = 25 + 49$ *calculate the squares*

$\qquad = 74$ *simplify*

So $h = \sqrt{74} = \mathbf{8.60\ cm}$ (2 dp).

Calculating the length of the diagonal of a rectangle, or the distance between two points on a co-ordinate grid, is the same as finding the hypotenuse in a right-angled triangle.

? To find a or b, just apply the rule as before, using $a^2 = h^2 - b^2$ or $b^2 = h^2 - a^2$.

C Finding sides with trigonometry

(1) The trigonometrical (trig for short) ratios for angle x in the triangle from section A are obtained by dividing pairs of sides.

The **sine of angle** $x = \sin x = \dfrac{\text{opposite}}{\text{hypotenuse}}$.

The **cosine of angle** $x = \cos x = \dfrac{\text{adjacent}}{\text{hypotenuse}}$.

The **tangent of angle** $x = \tan x = \dfrac{\text{opposite}}{\text{adjacent}}$.

(2) Use the 'phrase' SOH CAH TOA to remember these.

Suppose you had to find the side marked x in this triangle.

You are trying to find the side opposite to 32°, so underline the 'O's: S<u>O</u>H CAH T<u>O</u>A.

You know the length of the hypotenuse, so underline the 'H's: <u>S</u>O<u>H</u> CA<u>H</u> TOA.

<u>S</u>O<u>H</u> has two letters underlined, so this is a **sine** question.

$$\sin 32° = \frac{\text{opposite}}{\text{hypotenuse}} = \frac{x}{12}$$

$12 \times \sin 32° = x$ *multiply both sides by 12*

$$x = 6.36 \text{ cm (2 dp)}$$

(3) Sometimes the length to be found is on the bottom of the fraction:

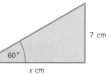

This problem involves the adjacent and opposite: S<u>O</u>H C<u>A</u>H T<u>OA</u>. Therefore you need **tangent** to solve it.

$$\tan 60° = \frac{\text{opposite}}{\text{hypotenuse}} = \frac{7}{x}, \text{ so } x \tan 60° = 7$$

$$x = \frac{7}{\tan 60°} = 4.04 \text{ cm (2 dp)}$$

D Finding angles with trigonometry

The inverse trig functions $\sin^{-1}$, $\cos^{-1}$ and $\tan^{-1}$ allow you to 'work backwards' from a sine, cosine or tangent to find the angle.

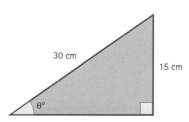

Find the angle θ in the diagram: <u>S</u>O<u>H</u> CA<u>H</u> TOA, so you need to use **sine**.

$$\sin \theta = \frac{\text{opposite}}{\text{hypotenuse}} = \frac{15}{30} = \textbf{0.5}$$

If you get a 'difficult' decimal when you divide, use the 'last answer' function on your calculator so you don't have to type it in again!

$$\theta = \sin^{-1}(0.5) = 30°.$$

>> practice questions

1 Find the sides marked with letters in the following triangles.

(a)   **(b)**   **(c)**   **(d)**

2 Find the missing angles in these triangles (not to scale):

(a)   **(b)**   **(c)**   **(d)**

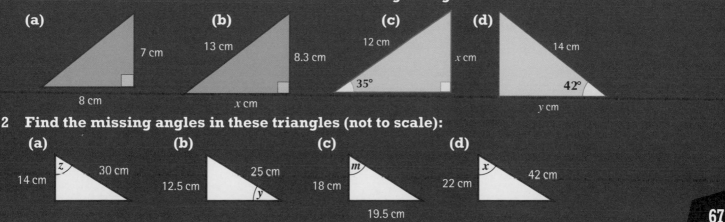

The sine rule

> The sine rule links the length of each side to the sine of the angle opposite it in the triangle.
>
> You can use trigonometry to find the area of a triangle.

A The sine rule

The standard ways of writing down the sine rule use the following notation for sides and angles in a general triangle.

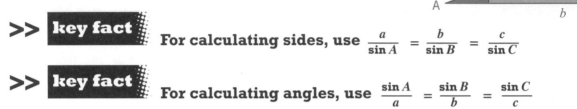

>> **key fact** For calculating sides, use $\dfrac{a}{\sin A} = \dfrac{b}{\sin B} = \dfrac{c}{\sin C}$

>> **key fact** For calculating angles, use $\dfrac{\sin A}{a} = \dfrac{\sin B}{b} = \dfrac{\sin C}{c}$

B Finding sides using the sine rule

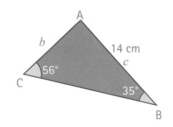

Find the length of the side marked b.

Using the sine rule, $\dfrac{c}{\sin C} = \dfrac{b}{\sin B}$

Now substitute into the formula: $\dfrac{14}{\sin 56°} = \dfrac{b}{\sin 35°}$

$b = \dfrac{14 \times \sin 35°}{\sin 56°} = 9.69\,\text{cm}$ (to 3 sf).

C Finding angles using the sine rule

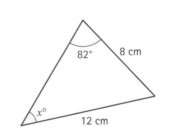

1 *Example*: Find the angle marked x.

Using the sine rule, $\dfrac{\sin x}{8} = \dfrac{\sin 82°}{12}$.

So $\sin x = 8 \times \dfrac{\sin 82°}{12} = 0.660\,17\ldots$

$x = \sin^{-1}(0.660\,17\ldots) = 41.3°$ (1 dp).

2 Notice that the sine of 138.7° is also 0.660 17…

In some cases, the unknown angle might turn out to be obtuse – check all available information to help you decide.

D The area of a triangle

1 The area of a triangle is given by $\frac{1}{2}ab \sin C$ (or $\frac{1}{2}bc \sin A$, or $\frac{1}{2}ac \sin B$).

2 *Example*: Find the area of this triangle.

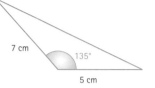

7 cm
135°
5 cm

Area $= \frac{1}{2}ab \sin C = \frac{1}{2} \times 5 \times 7 \sin 135° = 12.37\,cm^2$ (2 dp)

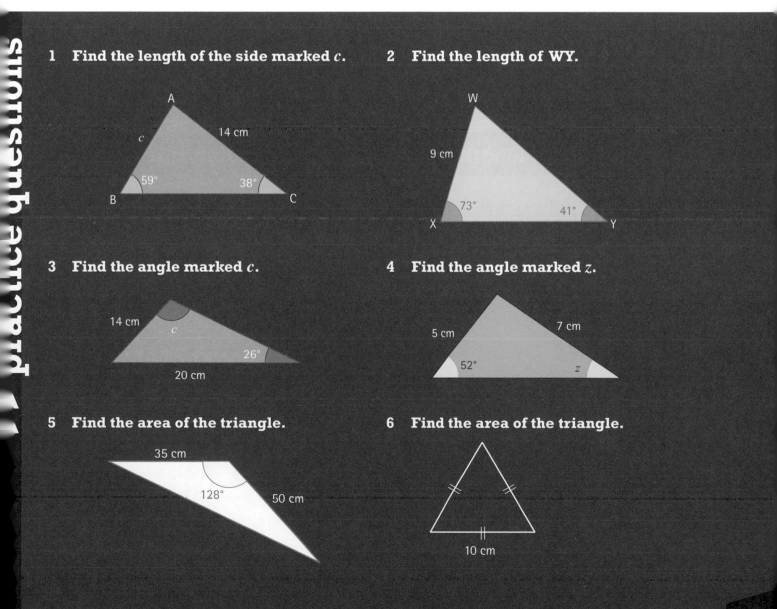

1 Find the length of the side marked c.

A
c
14 cm
59°
38°
B   C

2 Find the length of WY.

W
9 cm
73°
41°
X   Y

3 Find the angle marked c.

14 cm
c
26°
20 cm

4 Find the angle marked z.

5 cm
7 cm
52°
z

5 Find the area of the triangle.

35 cm
128°
50 cm

6 Find the area of the triangle.

10 cm

The cosine rule

> **Pythagoras' Rule** describes the link between the squares of the sides in a right-angled triangle. The cosine rule does the same in any triangle by introducing a new term involving the cosine of the angle opposite the side being calculated.

> The **sine and cosine rules** can be used in combination to solve triangles quickly and efficiently.

A   The cosine rule

Using the standard notation, the cosine rule states that:

>> **key fact** $a^2 = b^2 + c^2 - 2bc \cos A$.

As the vertices of the triangle can be named A, B and C any way you please, there are two other versions of the cosine rule:

$b^2 = c^2 + a^2 - 2ca \cos B$.

$c^2 = a^2 + b^2 - 2ab \cos C$.

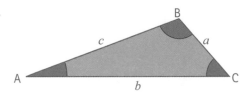

B   Finding the length of a side using the cosine rule

1   In triangle ABC, AC = 9 cm, BC = 7 cm, the angle at C is 42°. Find the length of AB.

2   As always, if they do not give you a diagram, make sure you draw one. Here is an example diagram:

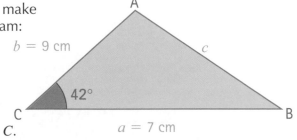

3   Using the cosine rule $c^2 = a^2 + b^2 - 2ab \cos C$.

4   Now substitute into this equation:

$c^2 = 7^2 + 9^2 - (2 \times 7 \times 9 \times \cos 42°)$

$c^2 = 36.36$

$c = 6.03$ cm (to 2 dp)

>> **key fact** When calculating an angle, rearrange

$a^2 = b^2 + c^2 - 2bc \cos A$, to read $\cos A = \dfrac{b^2 + c^2 - a^2}{2bc}$.

AB = 4 cm, BC = 6 cm, AC = 8 cm. Find the angle at A.

Using the cosine rule:

$\cos A = \dfrac{b^2 + c^2 - a^2}{2bc}$

$= \dfrac{8^2 + 4^2 - 6^2}{2 \times 8 \times 4}$

$\cos A = 0.6875$

Therefore $A = 46.57°$ (2 dp)

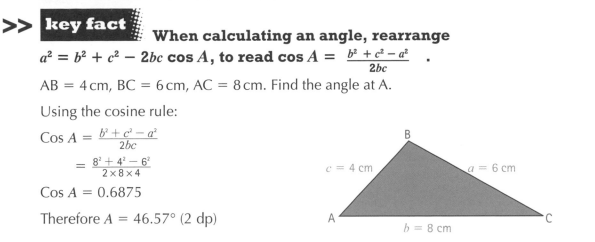

c = 4 cm   a = 6 cm   b = 8 cm

practice questions

Find the angles marked with letters:

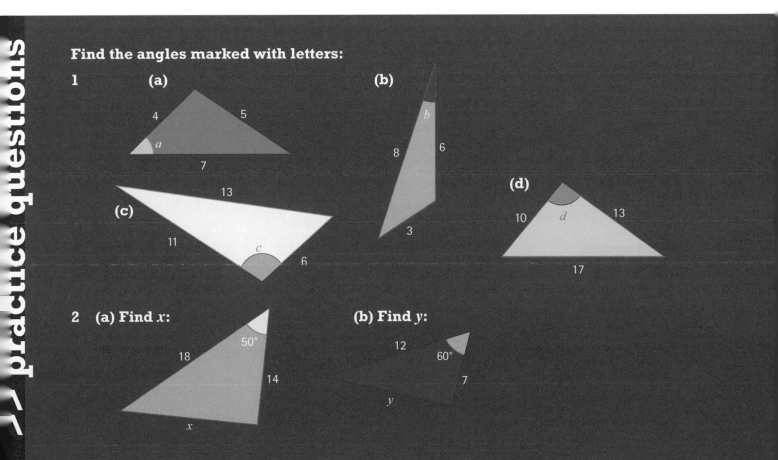

1 (a) (b)

(c) (d)

2 (a) Find *x*: (b) Find *y*:

3 A farmer creates a triangular sheep pen, MNP. MN is 40 m, NP is 35 m and the angle at N is 70°. Find the length of fence needed by the farmer.

4 A fishing trawler sets sail from harbour H and travels 40 km due North. This takes it to fishing grounds at F. After fishing for two hours, the trawler then heads on a bearing of 065° until it reaches new fishing grounds at G, a distance of 56 km. After fishing here, the trawler turns and heads back to harbour. Find the total distance travelled by the fishing trawler.

Trigonometric graphs: angles of any size

- Angles can be any size, positive or negative.

- The graphs of trigonometric functions repeat every 360° (sine and cosine) or 180° (tangent).

- Trigonometric equations may have several solutions.

A The sine of any angle

! The graph of $y = \sin x$ repeats every 360°. Its **period** is 360°.

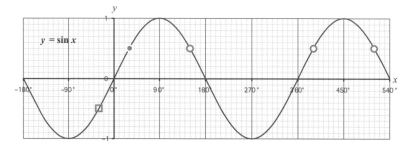

2 To solve the equation $\sin x = 0.5$, find $\sin^{-1}(0.5) = 30°$ (shown by the green dot on the graph). However, you can see that $180° - 30° = 150°$ is another solution (shown as a green circle). You can then add or subtract as many multiples of 360° as you like to find other solutions. So 390° and 510° are also solutions, as are $-210°$ and $-330°$, though these aren't shown on the graph. In solving a problem like this, you need to be given a **range** of angles to include.

3 If you were asked to solve $\sin x = -0.5$, you could start from $x = -30°$ (the green square) and use the same rules to find other solutions (210°, 330°, etc).

>> **key fact** Sin x is equal to sin $(360° + x)$ and also sin $(180° - x)$.
Sin $(-x) = -\sin x$.

B The cosine of any angle

! The period of $y = \cos x$ is also 360°.

2 You could be asked to solve an equation like $\cos x = \cos 78°$ $(-360° \leqslant x \leqslant 360°)$ on the non-calculator paper. From the graph, $\cos 78°$ is very close to 0.2. Another solution is $-78°$. You can then add or subtract multiples of 360° to obtain 282° and $-282°$.

So the solution can be written $x = \pm 78°$ or $\pm 282°$.

>> **key fact** Cos x is equal to cos $(360° + x)$ and also $(\cos 360° - x)$.
Cos $(-x) = \cos x$.

C The tangent of any angle

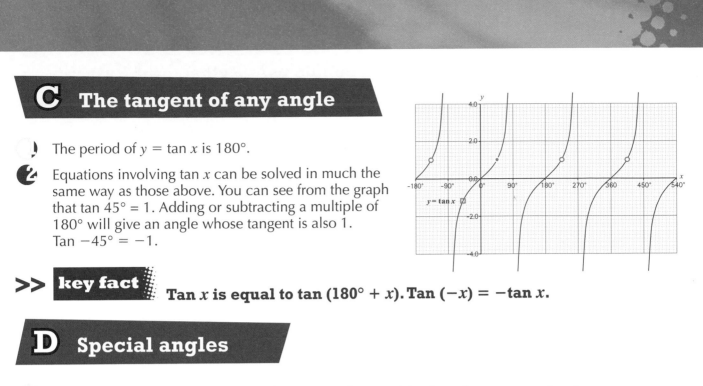

- The period of $y = \tan x$ is $180°$.

- Equations involving $\tan x$ can be solved in much the same way as those above. You can see from the graph that $\tan 45° = 1$. Adding or subtracting a multiple of $180°$ will give an angle whose tangent is also 1. $\tan -45° = -1$.

>> **key fact** Tan x is equal to tan $(180° + x)$. Tan $(-x) = -\tan x$.

D Special angles

- Most angles have trig functions with no particular special values, but some angles do.

  These diagrams illustrate them – check them with SOH CAH TOA.

Half an equilateral triangle

$\sin 30° = \cos 60° = \frac{1}{2}$

$\cos 30° = \sin 60° = \frac{\sqrt{3}}{2}$

$\tan 30° = \frac{1}{\sqrt{3}}$

$\tan 60° = \sqrt{3}$

Right-angled isosceles triangle

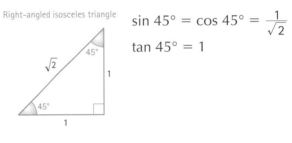

$\sin 45° = \cos 45° = \frac{1}{\sqrt{2}}$

$\tan 45° = 1$

>> **key fact** If x and y add up to $90°$, then $\sin x = \cos y$ and $\tan x \times \tan y = 1$.

- You should also be familiar with these, remembering that you can add multiples of the period of the graph:

  $\sin 0° = \sin 180° = 0$; $\sin 90° = 1$; $\sin -90° = -1$.

  $\cos 90° = \cos -90° = 0$; $\cos 0° = 1$; $\cos 180° = -1$.

  $\tan 0° = 0$; $\tan 90°$ is not defined.

>> practice questions

1 **Solve these equations correct to the nearest degree, within the ranges given.**

  (a) $\cos x = 0.9 \ (0° \leqslant x \leqslant 720°)$ (b) $\sin x = -0.75 \ (-360° \leqslant x \leqslant 360°)$

  (c) $\tan x = 3 \ (0° \leqslant x \leqslant 540°)$

2 **Solve these equations within the ranges given. Do not use a calculator.**

  (a) $\sin x = \sin 20° \ (0° \leqslant x \leqslant 720°)$ (b) $\tan x = \tan 100° \ (0° \leqslant x \leqslant 360°)$

  (c) $\cos x = \cos 88° \ (-360° \leqslant x \leqslant 360°)$

3 **Find exactly, leaving answers in surd form where necessary:**

  (a) $\sin 135°$ (b) $\tan 240°$ (c) $\cos 210°$ (d) $\sin 120°$

  (e) $\cos 315°$ (f) $\tan -150°$

Problems in three dimensions

- Finding the space diagonal of a cuboid uses Pythagoras' Rule in three dimensions.

- Trigonometry can be used in three-dimensional situations.

- It is usually best to split 3-D problems up into a number of 2-D diagrams.

A The space diagonal of a cuboid

The **space diagonal** of a cuboid goes from any vertex to the opposite vertex, through the centre of the cuboid. Suppose you have a cuboid with dimensions a, b and c:

The space diagonal is marked in red.

>> **key fact** The space diagonal of a cuboid
obeys Pythagoras' Rule in three dimensions: $x^2 = a^2 + b^2 + c^2$.

Find the space diagonal of a cube of side 10 cm.
For a cube,
$a = b = c$, so $x^2 = 10^2 + 10^2 + 10^2 = 300$.

$x = \sqrt{300} = 17.3$ cm to 3 sf

B Trigonometry in three dimensions

 There are many different ways that problems can be set. Here are a couple of examples. In each one, separate 'flat' diagrams are extracted from the main three-dimensional situation. This makes it clear what information to use, and allows you to concentrate on just one part of the problem at a time.

 ABCD is a regular tetrahedron. E is the centre of the base BCD, and AE is perpendicular to the base. Calculate angle EAC.

 The final answer will come from triangle EAC (on the right of the main diagram), but to be able to use this, you need to know the length of EC (marked e in the diagram). e can be calculated from the base, triangle BCD (below main diagram).

The midpoint of BC has been labelled F.

In triangle CEF, $\cos 30° = \dfrac{\text{adjacent}}{\text{hypotenuse}} = \dfrac{5}{e}$

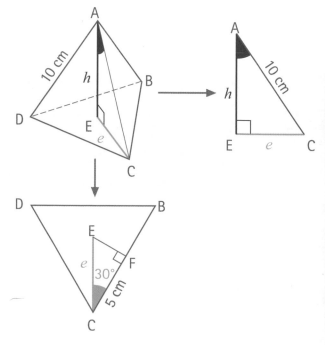

4 As $\cos 30° = \frac{\sqrt{3}}{2}$, $e = \frac{10}{\sqrt{3}}$. Leave this in surd form for the moment.

In triangle EAC, $\sin EAC = \frac{e}{10} = \frac{1}{\sqrt{3}}$. So angle $EAC = \sin^{-1}\left(\frac{1}{\sqrt{3}}\right) = 35.3°$ to 1 dp.

5 In the cuboid shown, solve triangle ACH.

First, find the lengths (a, b, c) of the sides,
using Pythagoras' Rule on the faces of the cuboid.

This gives $\quad a = \sqrt{449} = 21.2$ cm to 1 dp,

$\qquad\qquad b = \sqrt{425} = 20.6$ cm to 1 dp,

$\qquad\qquad c = \sqrt{74} = 8.6$ cm to 1 dp.

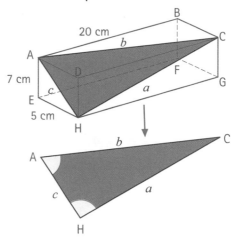

6 Using the diagram above, the cosine rule states that $a^2 = b^2 + c^2 - 2bc\cos A$,

so that $A = \cos^{-1}\left(\frac{b^2 + c^2 + a^2}{2bc}\right) = \cos^{-1}\left(\frac{425 + 74 + 449}{2 \times \sqrt{425 \times 74}}\right) = 81.9°$ to 1 dp.

The sine rule then states that $\frac{\sin C}{c} = \frac{\sin A}{a}$, so that

$C = \sin^{-1}\left(\frac{c \sin A}{a}\right) = \sin^{-1}\left(\frac{449 \sin 81.995...°}{\sqrt{74}}\right) = 23.7°$ to 1 dp.

$H = 180° - A - C = 74.4°$ to 1 dp.

All the sides and angles have now been found.

>> practice questions

1 Calculate the space diagonal of:

(a) a cuboid 2 cm by 3 cm by 4 cm.

(b) a cuboid 10 cm by 5 cm by 20 cm.

(c) a unit cube.

2 John made a square-based pyramid out of wire. The perpendicular height of his model was 20 cm and the base was 10 cm on a side. Assuming that he only constructed the 8 edges of the pyramid, what length of wire did John use?

3 The diagram shows two equally tall observers, A and B, looking at a transmitter mast CM which is 10 m taller than they are. A is due south of C, and the angle of elevation of M from her position is 16°. B is due east of C and registers an angle of elevation of 32°. What is the bearing of B from A?

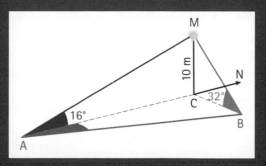

Calculating areas

- Area is the amount of surface covered by a 2-dimensional shape.

- Most simple shapes have formulae that can be used to calculate their areas.

- More complicated (compound) shapes can be built from simple ones.

A Rectangles and squares

>> **key fact** The area of a rectangle = length × width (*lw*).
The area of a square = length × length (l^2).

1 In this rectangle $l = 100$ and $w = 25$.

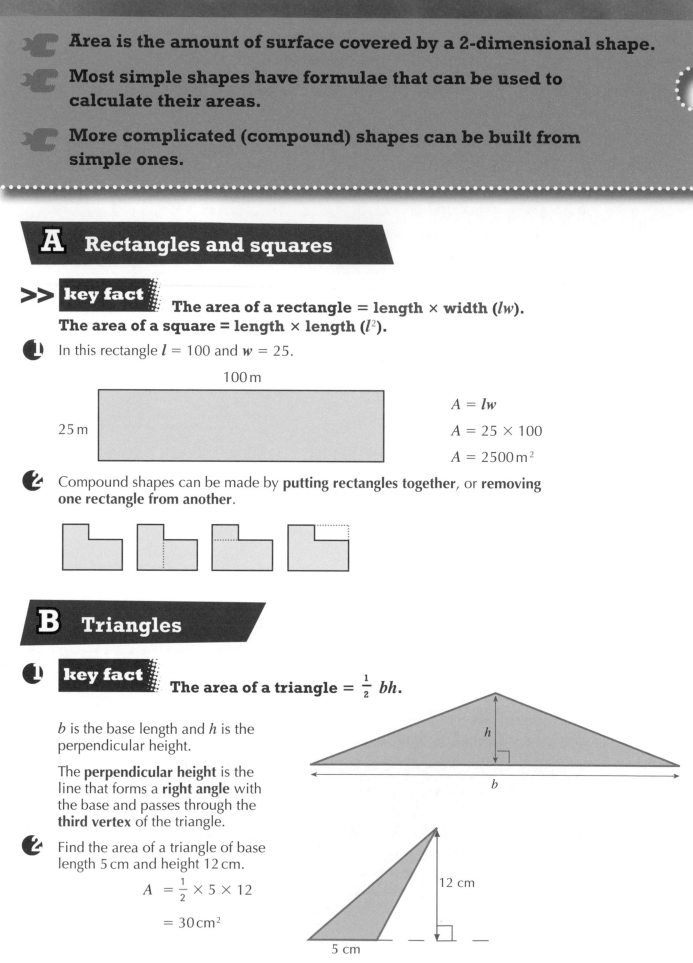

100 m

25 m

$A = lw$

$A = 25 \times 100$

$A = 2500\,\text{m}^2$

2 Compound shapes can be made by **putting rectangles together**, or **removing one rectangle from another**.

B Triangles

1 **key fact** The area of a triangle = $\frac{1}{2} bh$.

b is the base length and *h* is the perpendicular height.

The **perpendicular height** is the line that forms a **right angle** with the base and passes through the **third vertex** of the triangle.

h

b

2 Find the area of a triangle of base length 5 cm and height 12 cm.

$A = \frac{1}{2} \times 5 \times 12$

$= 30\,\text{cm}^2$

12 cm

5 cm

C Parallelograms and trapezia

1 Parallelograms and trapezia both use the perpendicular height in their area formulae:

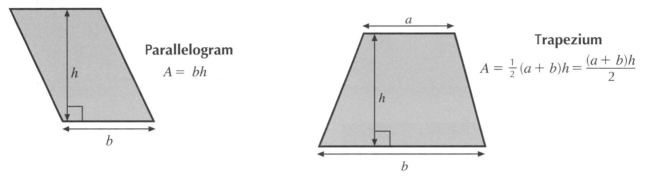

Parallelogram

$A = bh$

Trapezium

$A = \frac{1}{2}(a + b)h = \frac{(a + b)h}{2}$

2 Calculate the area of this trapezium:

The perpendicular height isn't given, so you will have to calculate it using Pythagoras' Rule and the triangle on the right.

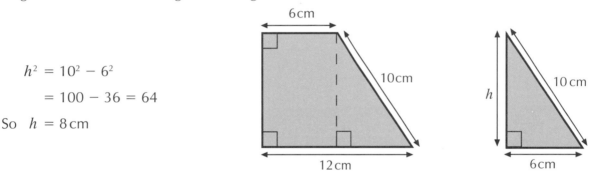

$$h^2 = 10^2 - 6^2$$
$$= 100 - 36 = 64$$

So $h = 8\,cm$

Using the trapezium formula with $a = 6\,cm$, $b = 12\,cm$, and $h = 8\,cm$ gives:

$A = \frac{1}{2}(a + b)h = 72\,cm^2$.

>> practice questions

1 Find the area of this rectangle.

5m

3.5m

2 Find the area of a carpet needed for this office layout.

10 m

11m

23 m

8m

3 Catherine has to paper a bedroom wall that is 3 m high and 10 m long. The wallpaper she wants comes in rolls 2 m wide and 5 m long. How many rolls of paper will she need?

4 Martin wants to paint a fence. The fence is 14 m long and 3 m high. He has to paint both sides of the fence. One tin of paint covers 12 m² of fence. How many tins will he need?

5 Find the area of a triangle that is 6.5 cm in width and 7 cm in height.

6 A trapezium has parallel sides of length 5 cm and 11 cm. Its area is 80 cm². What is its perpendicular height?

Circle calculations

- The special number π (pi) features in calculations involving circles. π ≈ 3.142

- The circumference of a circle is given by π × diameter.

- The area of a circle is given by π × (radius)².

A Parts of a circle

) The diagrams give the names of parts of a circle.

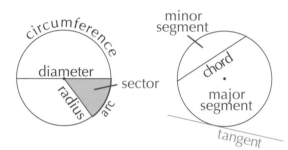

B Circumference calculations

) The length of the circumference of a circle is just over 3 times its diameter. The actual number is called **pi** (written π, pronounced 'pie'). You should have a π key on your calculator. If not, use an approximate decimal value such as 3.14 or 3.142.

> **key fact** Circumference = π × diameter, or $C = \pi d$.

So if a circle is 5 cm across, its circumference is $C = \pi d = \pi \times 5 = 15.71$ cm (2 dp).

) If you are given the radius, you can double it to find the diameter and then use $C = \pi d$. Alternatively, you can use the other circumference formula:

>> **key fact** Circumference = 2 × π × radius, or $C = 2\pi r$.

So if a circle's radius is 10 m, its circumference is $C = 2\pi r = 2 \times \pi \times 10 = 62.83$ m (2 dp).

> **key fact** To work backwards from the circumference to find the diameter, use $d = \frac{C}{\pi}$.

Example:

A trundle wheel's circumference is exactly 1 m. What is its diameter?

$d = \frac{1}{\pi} = 0.318$ m $= 31.8$ cm (3 sf).

C Area calculations

1 **key fact** Area of a circle = π × the square of the radius, or $A = \pi r^2$.

The circle from section B2 with diameter 5 cm has radius 2.5 cm.

So its area is
$A = \pi r^2 = \pi \times 2.5^2 = 19.63 \text{ cm}^2$ (2 dp).

2 Working backwards from the area to the radius is a little harder: $r = \sqrt{\frac{A}{\pi}}$.

So a circular pond covering an area of 4 m² would have radius $r = \sqrt{\frac{4}{\pi}} = 1.13$ m.

D Leaving π in the answer

1 In some exam questions, you may be asked to write an answer 'in terms of π'. This means that, instead of calculating a decimal answer using the π button on your calculator, you simply write down what multiple of π is involved. You can then easily compare two answers.

2 *Example*:

Circle *A* has diameter 20 cm. Circle *B* has radius 20 cm. Write the areas of the two circles in terms of π, and find how many times larger the area of *B* is than that of *A*.

Radius of $A = 20 \div 2 = 10$ cm.

Area of $A = \pi \times 10^2 = 100\pi$.
Area of $B = \pi \times 20^2 = 400\pi$.

The area of *A* is 4 times the area of *B*.

E Arcs and sectors

To find the length of an arc or the area of a sector, calculate the circumference or area of the complete circle, and find the correct fraction of this.

Example: Find the perimeter and area of this shape:

The diameter of the circle is 24 cm. The length of the arc is $\frac{80}{360}$ of the circumference.

Perimeter = $(\frac{80}{360} \times \pi \times 24) + (2 \times 12) = 16.7551\ldots + 24 = 40.76$ cm (2 dp).

Area = $\frac{80}{360} \times \pi \times 12^2 = 100.53$ cm² (2 dp).

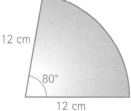

12 cm

80°

12 cm

>> practice questions

1 Find the circumference and area of each of the following circles, using the value for π on your calculator:

   (a) radius 7 cm   **(b)** diameter 4 cm   **(c)** diameter 25 cm   **(d)** radius 10.9 cm

2 A circular pond has a diameter of 4.5 m. It is surrounded by a path that is 1.2 m wide.

   Find:

   (a) the area of the pond

   (b) the area of the path.

3 A circle has an area of 100 cm². Find the radius.

Volume calculations

- A cuboid is a prism with a rectangular cross-section, so its **volume = length × width × height**.

- The volume of any prism is equal to the area of its cross-section multiplied by its length (or height).

- A cylinder is a circular prism.

A Cuboids

Volume is the amount of three-dimensional space taken up by a solid shape.

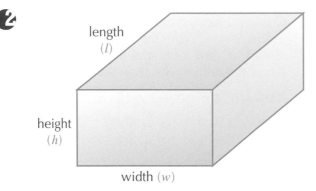

length (*l*)

height (*h*)

width (*w*)

>> **key fact** Volume of a cuboid = length × width × height; $V = lwh$.

Example:

A room is in the shape of a cuboid 8 m long, 4.5 m wide and 3 m high. What is its volume?

$$V = lwh = 8 \times 4.5 \times 3 = 108\,m^3.$$

Remember: volume is measured in cubic units.

B Prisms

area of end or cross-section (*A*)

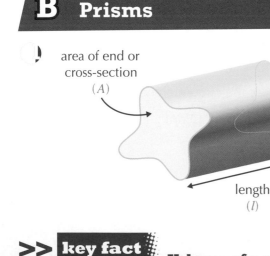

length (*l*)

>> **key fact** Volume of a prism = ⌐on × length; $V = Al.$

'standing on end', the ⌐ay be thought of as a ⌐a would be $V = Ah$.

Example:

Find the volume of this triangular prism.

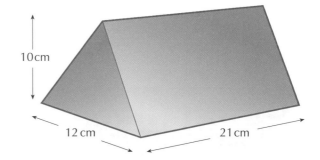

10 cm

12 cm   21 cm

Area of cross section

$$A = \tfrac{1}{2}bh = \tfrac{1}{2} \times 12 \times 10 = 60\,cm^2.$$

Volume $V = Al = 60 \times 21 = 1260\,cm^3.$

C Cylinders

key fact Cylinders are prisms with circular cross-sections. The volume formula is $V = Al = \pi r^2 l$.

Example:

Find the capacity of this cylindrical can.

4 cm

10 cm

Area of cross-section
$A = \pi r^2 = \pi \times 4^2 = 16\pi$.

Volume of cylinder $V = Ah = 16\pi \times 10$
$= 503 \text{ cm}^3$ (to nearest cm³).

As $1 \text{ cm}^3 = 1 \text{ ml}$, the can will hold 503 ml or 0.503 l (to 3 sf).

Note that, if required, the answer could have been left in terms of π: 160π ml.

D Pyramids

A pyramid has a flat base joined to an apex. The perpendicular height joins the apex to the base at right angles (like the height in a triangle).

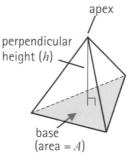

apex

perpendicular height (h)

base (area = A)

key fact The volume of a pyramid is $V = \frac{1}{3}Ah$.

key fact Cones are pyramids with circular bases. The volume formula is $V = \frac{1}{3}Ah = \frac{1}{3}\pi r^2 h$.

Example: A conical water cup is 11 cm high and the diameter of the rim is 6 cm. Find its capacity, correct to the nearest centilitre.

The 'base' of this cone is the circle made by the rim. Its radius is 3 cm.

$V = \frac{1}{3} \times \pi \times 3^2 \times 11 = 33\pi = 103 \text{ cm}^3 = 103 \text{ ml} = 10.3 \text{ cl} = 10 \text{ cl}$ (nearest cl).

E Spheres

key fact The volume of a sphere of radius r is $V = \frac{4}{3}\pi r^3$.

>> practice questions

1 Find the volume of a cuboid of dimensions 2.5 m, 4 m and 6 m.

2 A swimming pool is half full of water. If the dimensions of the pool are 9 m by 12 m by 2 m, how many cubic metres of water are in the pool?

3 Find the volume of a cylinder that has a base radius of 10 cm and a height of 12 cm.

4 Julie has a waste bin that is cylindrical in shape. The base has a radius of 0.49 m and the height is 0.82 m. Find the volume of the bin.

5 A kart-racing track has a tunnel shaped like a triangular prism. The dimensions of the tunnel are: 13 m long, 3 m wide and 4 m from its highest point to the ground. Find the volume of the tunnel.

6 A time capsule in the shape of a cylinder is being placed in a wall for 25 years. The dimensions of the cylinder are: base radius 12 cm, length 100 cm. The hole in the wall has to have 10% more volume than the cylinder. What will the volume of the hole be?

Congruence and similarity

- Congruent 2-D shapes are shapes that are identical.

- Shapes that are mathematically similar are identical, except in size.

- In exam questions, congruent and similar shapes are not usually drawn accurately.

A Congruent triangles

1. There are certain tests for triangles to establish if they are congruent.

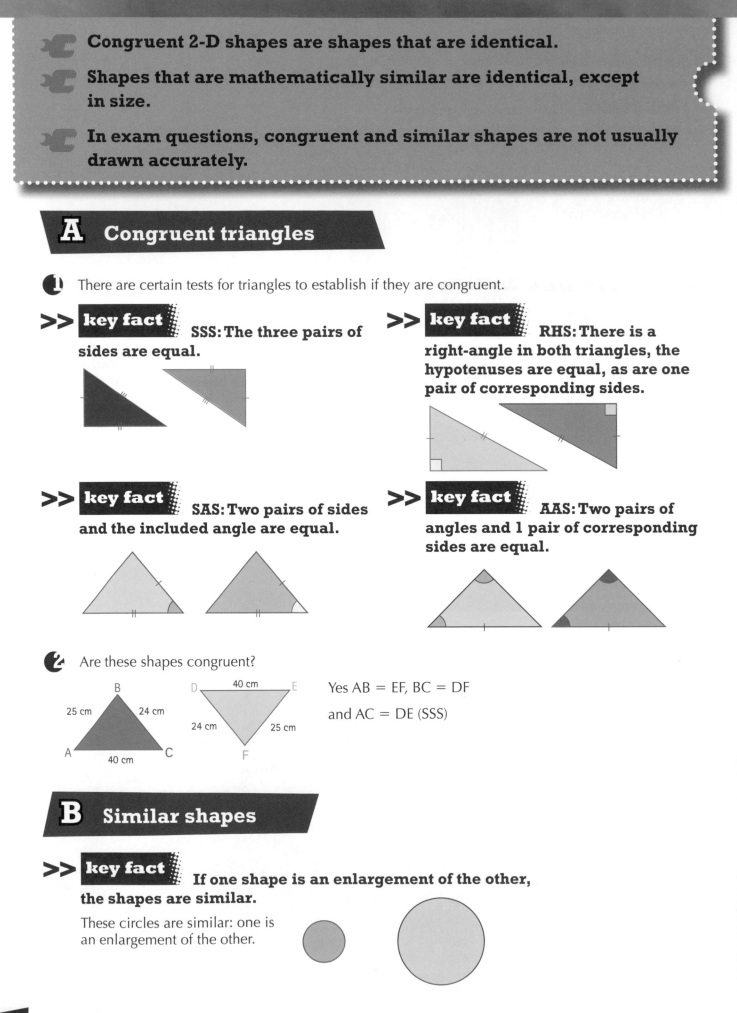

>> **key fact** SSS: The three pairs of sides are equal.

>> **key fact** RHS: There is a right-angle in both triangles, the hypotenuses are equal, as are one pair of corresponding sides.

>> **key fact** SAS: Two pairs of sides and the included angle are equal.

>> **key fact** AAS: Two pairs of angles and 1 pair of corresponding sides are equal.

2. Are these shapes congruent?

B
25 cm   24 cm
A 40 cm C

D ___ 40 cm ___ E
24 cm   25 cm
F

Yes AB = EF, BC = DF

and AC = DE (SSS)

B Similar shapes

>> **key fact** If one shape is an enlargement of the other, the shapes are similar.

These circles are similar: one is an enlargement of the other.

C The scale factor of an enlargement

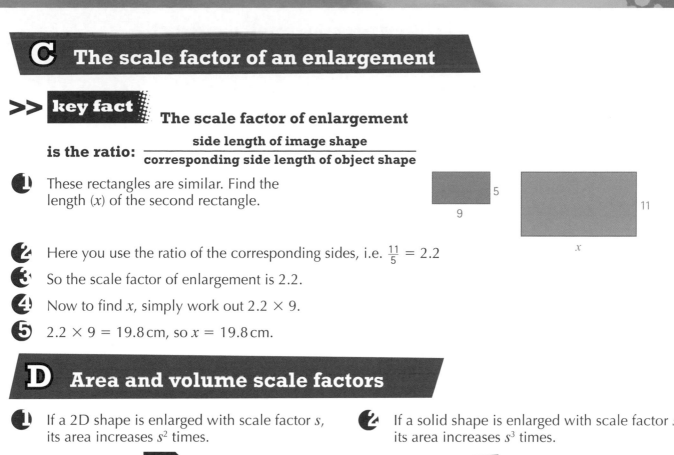

>> key fact

The scale factor of enlargement

is the ratio: $\dfrac{\text{side length of image shape}}{\text{corresponding side length of object shape}}$

① These rectangles are similar. Find the length (x) of the second rectangle.

② Here you use the ratio of the corresponding sides, i.e. $\frac{11}{5} = 2.2$

③ So the scale factor of enlargement is 2.2.

④ Now to find x, simply work out 2.2×9.

⑤ $2.2 \times 9 = 19.8\,\text{cm}$, so $x = 19.8\,\text{cm}$.

D Area and volume scale factors

① If a 2D shape is enlarged with scale factor s, its area increases s^2 times.

The scale factor for this enlargement is 3: the area of the image is $3^2 = 9$ times bigger.

This also applies to surface area for solid shapes.

② If a solid shape is enlarged with scale factor s, its area increases s^3 times.

The scale factor for this enlargement is 5: the area of the image is $5^3 = 125$ times bigger.

>> practice questions

1 Explain why these triangles must be congruent.

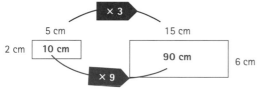

2 These rectangles are similar: find the length marked a. What is the area scale factor of the enlargement?

3 These shapes are similar. Find the unknown lengths. What is the volume scale factor of the enlargement?

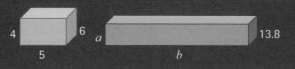

4 Are these L-shapes congruent, similar or neither? Explain your answer.

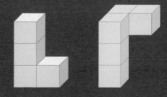

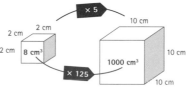

Constructions

> ✎ **When you are asked to 'draw accurately' or 'construct' a diagram, you will need the techniques on this double page.**
>
> ✎ **The set-square in your mathematical instrument kit is not for drawing right angles!**

A Triangles

❶ To draw a triangle accurately given all three sides:
for example, 8 cm, 5 cm and 4 cm:

Start with a base line, 8 cm long.	*Set your compass to 5 cm. Put the point on A and draw an arc where you think the other vertex will be.*	*Set your compass to 4 cm. Put the point on B and draw another arc, crossing the first.*	*Join A and B to the point where the arcs cross.*

❷ To draw a triangle given one side and two angles:
for example, 10 cm, with 50° at one end and 30° at the other:

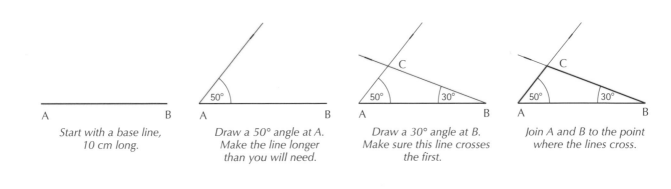

Start with a base line, 10 cm long.	*Draw a 50° angle at A. Make the line longer than you will need.*	*Draw a 30° angle at B. Make sure this line crosses the first.*	*Join A and B to the point where the lines cross.*

❸ You can deal with other combinations of sides and angles by mixing the two methods.

>> **key fact** To construct a 60° angle, just construct an equilateral triangle.

B Perpendicular bisectors

❶ **key fact** A line that passes through the midpoint of two given points and is perpendicular to the line joining the points is called their **perpendicular bisector**.

84

To construct a perpendicular bisector:

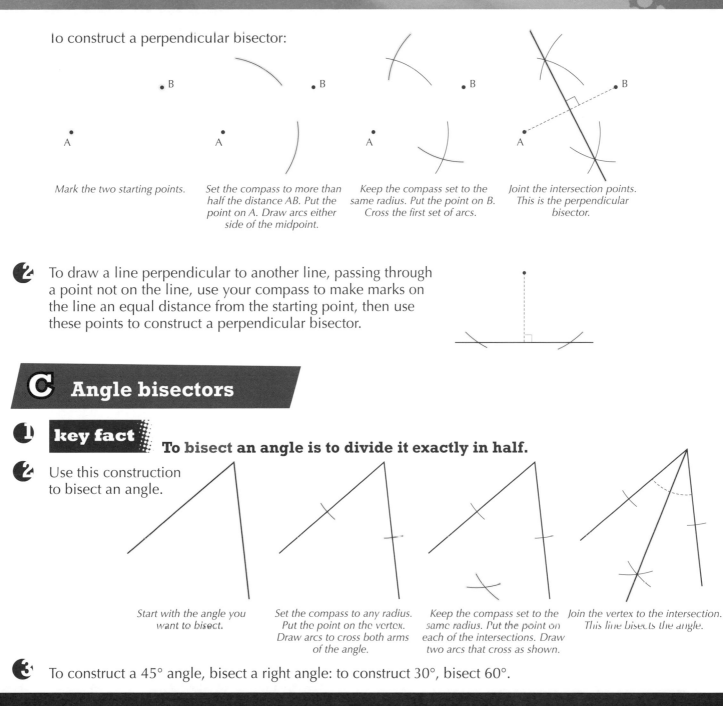

Mark the two starting points.

Set the compass to more than half the distance AB. Put the point on A. Draw arcs either side of the midpoint.

Keep the compass set to the same radius. Put the point on B. Cross the first set of arcs.

Joint the intersection points. This is the perpendicular bisector.

2 To draw a line perpendicular to another line, passing through a point not on the line, use your compass to make marks on the line an equal distance from the starting point, then use these points to construct a perpendicular bisector.

C Angle bisectors

1 **key fact** To bisect an angle is to divide it exactly in half.

2 Use this construction to bisect an angle.

Start with the angle you want to bisect.

Set the compass to any radius. Put the point on the vertex. Draw arcs to cross both arms of the angle.

Keep the compass set to the same radius. Put the point on each of the intersections. Draw two arcs that cross as shown.

Join the vertex to the intersection. This line bisects the angle.

3 To construct a 45° angle, bisect a right angle: to construct 30°, bisect 60°.

>> practice questions

1 (a) **Construct a triangle with sides of 10 cm, 8 cm and 6 cm.**
   (b) **Construct the perpendicular bisectors of all three sides. They should cross at a single point.**
   (c) **Put the point of your compass on the intersection and the pencil on one of the vertices. Draw a complete circle – it should pass through all three vertices.**

2 (a) **Construct a triangle with a 7 cm side, a 70° angle at one end and a 45° angle at the other.**
   (b) **Bisect all three angles. The bisectors should cross at a single point.**
   (c) **Put the point of your compass on the intersection. Adjust your compass and draw a circle that just touches all three sides of the triangle.**

3 **Using only your compass and ruler, construct angles of 60° and 90° next to each other to make an angle of 150°. Bisect this angle. Use your protractor to check that each half is 75°.**

Loci

- A locus is a set of positions produced by a mathematical rule.

- Most loci can be drawn using the constructions on pages 84–85.

- Usually you will need a combination of different loci to answer a question.

A What are loci?

1 **key fact** A locus is a set of positions generated by some rule.

Different loci are generated by different rules. A locus may be:

- part of a **line** or **curve**

- a **region** (area), inside or outside a shape.

2 Exam questions often depend on combining more than one rule to produce a locus.

B Fixed distances

1 The rule 'stay a fixed distance away from a fixed point' gives a circular locus.

2 The locus of a point that stays a fixed distance from a straight line is two parallel straight lines.

3 You may need to combine these two locus types.

This locus is '2 cm from the line segment':

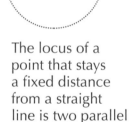

This locus is 'within 2 cm of the square'.

C Equal distances

1 **key fact** A position that is the same distance from two objects is equidistant from them.

The rule 'equidistant from two fixed points' is a straight line – the **perpendicular bisector** of the two points.

Any point on this line is equidistant from P and Q.

2 The rule 'equidistant from two straight lines' is a straight line – the **angle bisector** of the two lines.

Any point on this line is equidistant from XY and YZ.

These loci are 'closer to P than to Q'

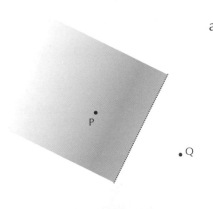

and 'closer to XY than to XZ'.

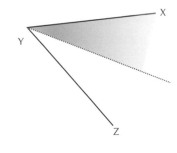

D Combining loci

Questions about combined loci usually involve:

- drawing two or more loci

- finding the points that are part of all the loci you have drawn.

The diagrams show how to answer the following question:

Point A is 4 cm from O. Indicate on your diagram the region containing points that are within 4 cm of O, and also closer to A than to O.

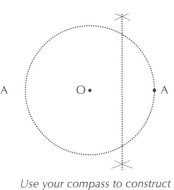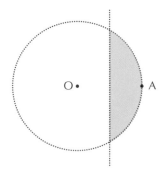

Mark O and A, 4 cm apart.

Set the compass to 4 cm. Draw a circle centred on O. It should pass through A.

Use your compass to construct the perpendicular bisector for O and A.

Shade the region inside the circle and to the right of the bisector.

>> practice questions

1 (a) **Three radio transmitters R, S and T are positioned so that R is 90 km from S and 60 km from T; also, S is 75 km from T. Using a scale of 1 cm : 10 km, draw a diagram to show the positions of the transmitters.**

   (b) **Broadcasts from R can be received up to 50 km away, from S up to 60 km away, and from T up to 35 km away. Indicate on your diagram the reception region for each broadcast. Shade in the region where all three broadcasts can be heard.**

   (c) **How close could you be to S and still receive all three broadcasts?**

2 (a) **Draw an equilateral triangle with sides 7 cm long.**

   (b) **Construct the locus of points that are exactly 1.5 cm from the triangle.**

3 **Draw a co-ordinate grid with x and y scales going from 0 to 10.**

   (a) **Draw the locus of points that are equidistant from points (9, 2) and (9, 8).**

   What is the equation of this line?

   (b) **Draw the locus of points that are equidistant from points (2, 4) and (6, 8).**

   What is the equation of this line?

   (c) **Which point lies on both loci?**

Vectors

- Vectors cannot be multiplied or divided by each other, but can be multiplied by numbers called **scalars**.

- The normal rules of algebra apply to vectors and scalars.

- The magnitude (length) of a vector can be determined using Pythagoras' Rule.

A Multiplying vectors by scalars

1 Suppose vector **c** is $\binom{3}{4}$. Then $\mathbf{c} + \mathbf{c} = \binom{3}{4} + \binom{3}{4} = \binom{6}{8}$.

This vector can also be written as 2**c**. 2 is an ordinary number with size but no direction. These numbers are sometimes called **scalars**, to distinguish them from vectors, that have a direction.

>> **key fact** When a vector is multiplied by a scalar, its length changes but its direction remains the same, unless the scalar is negative, when the direction is reversed.

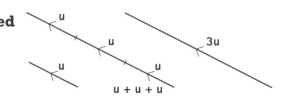

2 To multiply a vector in column form by a scalar, simply multiply both components.

$2\mathbf{c} = \binom{2 \times 3}{2 \times 4} = \binom{6}{8}$

B Vector algebra

1 The normal rules of algebra apply to vectors and scalars, so you can simplify expressions containing them.

In the diagram, $\overrightarrow{AB} = \mathbf{p} + 3\mathbf{q} + 2\mathbf{p} + \mathbf{q} - 4\mathbf{p}$. Collecting like terms, this simplifies to

$\overrightarrow{AB} = -\mathbf{p} + 4\mathbf{q}$ or $4\mathbf{q} - \mathbf{p}$.

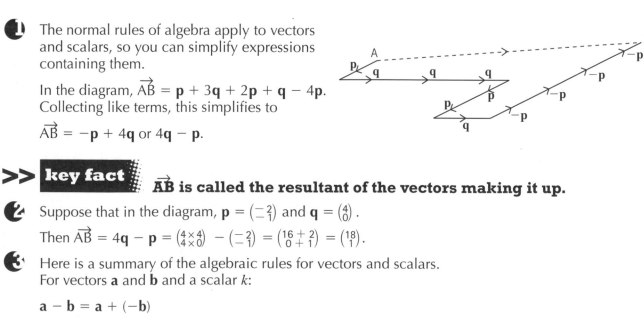

>> **key fact** $\overrightarrow{AB}$ is called the resultant of the vectors making it up.

2 Suppose that in the diagram, $\mathbf{p} = \binom{-2}{-1}$ and $\mathbf{q} = \binom{4}{0}$.

Then $\overrightarrow{AB} = 4\mathbf{q} - \mathbf{p} = \binom{4 \times 4}{4 \times 0} - \binom{-2}{-1} = \binom{16 + 2}{0 + 1} = \binom{18}{1}$.

3 Here is a summary of the algebraic rules for vectors and scalars.
For vectors **a** and **b** and a scalar k:

$\mathbf{a} - \mathbf{b} = \mathbf{a} + (-\mathbf{b})$

$k\mathbf{a} = \mathbf{a} + \mathbf{a} + \ldots + \mathbf{a} + \mathbf{a}$ (k times)

$k(\mathbf{a} + \mathbf{b}) = k\mathbf{a} + k\mathbf{b}$

C Using position vectors

There are two particular uses of position vectors that you need to be aware of.

key fact If points A and B have position vectors **a** and **b**, respectively, then $\overrightarrow{AB} = \mathbf{b} - \mathbf{a}$.

If A is the point (4, 8) and B is (6, 3), then $\overrightarrow{AB} = \mathbf{b} - \mathbf{a} = \binom{6}{3} - \binom{4}{8} = \binom{2}{-5}$.

key fact If points A and B have position vectors **a** and **b**, respectively, then the midpoint M of AB has position vector $\mathbf{m} = \frac{1}{2}(\mathbf{a} + \mathbf{b})$.

Using points A and B from step 1 above, the midpoint of AB has position vector

$\frac{1}{2}(\mathbf{a} + \mathbf{b}) = \frac{1}{2}\binom{4}{8} + \binom{6}{3} = \frac{1}{2}\binom{10}{11} = \binom{5}{5\frac{1}{2}}$. So the midpoint is located at $(5, 5\frac{1}{2})$.

D The magnitude of a vector

Any vector not parallel to the co-ordinate axes can be thought of as the hypotenuse of a right-angled triangle. Pythagoras' Rule gives its length.

The length of a vector x is called its magnitude or modulus and is written $|x|$.

>> key fact The magnitude of a vector $\binom{a}{b} = \sqrt{a^2 + b^2}$.

For example, the size of the vector $\binom{3}{-4}$ is $\sqrt{3^2 + (-4)^2} = \sqrt{9 + 16} = \sqrt{25} = 5$ units.

>> practice questions

1 In this question, $\mathbf{a} = \binom{1}{5}$, $\mathbf{b} = \binom{7}{0}$, $\mathbf{c} = \binom{-2}{2}$, $\mathbf{d} = \binom{0}{-3}$, $n = 3$, $m = -2$.

For each part, (i) simplify the expression if possible, (ii) write the resultant as a column vector, (iii) find its magnitude, correct to 2 dp where necessary.

(a) $\mathbf{a} + \mathbf{b}$

(b) $4\mathbf{c}$

(c) $2\mathbf{d} + \mathbf{b} - \mathbf{d}$

(d) $n\mathbf{a} + m\mathbf{a}$

(e) $\mathbf{a} + 3\mathbf{d} - \mathbf{c} - 2\mathbf{a}$

2 On a co-ordinate grid, point J is at (2, 5) and K is at (6, −1). Find as column vectors:

(a) the position vector of J

(b) the position vector of K

(c) $\overrightarrow{JK}$

(d) $\overrightarrow{KJ}$

(e) the position vector of the midpoint of JK

Transformations

- A transformation changes shapes by altering their position or size.

- The original shape is called the object. A transformation produces the image shape.

- For enlargements, images are similar to their objects. For translations, reflections and rotations, the images are congruent to their objects.

A Translation

① key fact Translations are sliding movements. All they do is change the position of an object.

Object triangle A in the diagram has been translated 5 units to the right and 4 units down to give image triangle B.

② Translations can be described using **column vectors**. The vector for the translation shown is $\begin{pmatrix} 5 \\ -4 \end{pmatrix}$. Positive numbers are used for movements up or right, negative for down or left.

B Reflection

To describe a reflection, you need to specify where the mirror line is. The only mirror lines you will be asked to use are horizontal, vertical or at 45° to the axes.

>> key fact Points in the object are the same distance from the mirror as their images.

② The diagram shows the object triangle, P, reflected in the line $x = 4$ to produce image Q. The point (2, 1) of P is 2 units away from the mirror; so is its image in Q, (6, 1).

③ In the diagram, P is also reflected in the line $y = x$ to produce image R. (2, 1) becomes (1, 2) in R: they are both half a *diagonal* unit away from the mirror.

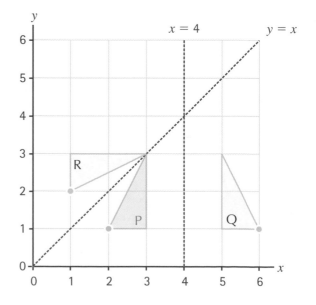

C Rotation

1 **key fact** To describe a rotation, you need the angle the object turns through, and a fixed position called the **centre of rotation**.

2 The only angles used in exam questions are 90° (clockwise or anticlockwise) and 180°. The diagram shows object shape U rotated by 180° about centre (1, 3) to give image V. It also shows U rotated 90° clockwise about centre (4, 1) to give image W.

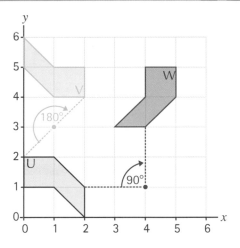

D Enlargement

1 To describe an enlargement, you need the scale factor, and a fixed position called the **centre of enlargement**.

>> key fact The scale factor controls how many times larger than the object the image is. Enlargements are mathematically **similar** to their objects.

2 The diagram shows the object shape A enlarged using the origin as the centre. Image B was enlarged with scale factor 2, image C with scale factor 3.

3 Scale factors smaller than 1 produce an image smaller than the object. When the scale factor is negative, the object and image are on opposite sides of the centre of enlargement.

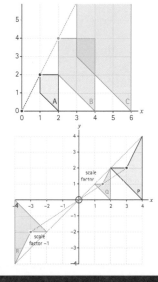

>> practice questions

For these questions, you will need to draw a co-ordinate grid with x- and y- axes from −10 to 10. Draw the object shape: a triangle with vertices (4, 0), (4, 2) and (3, 2).

1 For each part, write down the co-ordinates of the vertices of the image.

(a) Translate using the given vectors:

(i) $\begin{pmatrix} 0 \\ 6 \end{pmatrix}$;   (ii) $\begin{pmatrix} 5 \\ 3 \end{pmatrix}$.

(b) Reflect in the following lines:

(i) $x = 5$;   (ii) $y = x$.

(c) Rotate as follows:

(i) 90° clockwise, centre (−3, 4);
(ii) 180°, centre (2, −2).

(d) Enlarge with centre (6, 5):

(i) scale factor 2;
(ii) scale factor −1.

2 Describe fully the transformation that maps the vertices of the object to:

(a) (1, 0), (1, 2), (0, 2)

(b) (4, −6), (4, −8), (3, −8)

(c) (−1, 7), (−3, 7), (−3, 6)

(d) (−5, 0), (−5, 8), (−9, 8)

Pie charts

Pie charts are useful when you want to see what fraction of a whole each item in a survey represents.

A What is a pie chart?

>> **key fact** Pie charts are circular diagrams – the shape of a pie – split up to show sectors, or 'slices', that represent each part of the survey. By looking at the size of each sector, it is possible to estimate the fraction of the total for each data item.

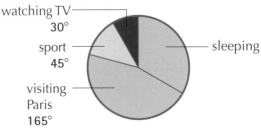

This pie chart shows how a group of students on a school trip in France spent the day and represents a full 24-hour day.

2 They had to travel a long way from their hostel to visit Paris. What fraction of the day was spent sleeping?

To find the answer, you have to work out the angle for the sleeping sector.

A pie chart is a circle and there are 360° in a circle.

Add up the angles for the other sectors.

$30° + 45° + 165° = 240°$

So, the angle for the sleeping sector is:

$360° − 240° = 120°$

The fraction for sleeping is $\frac{120°}{360°}$, which cancels to $\frac{1}{3}$.

3 How many hours in the day were spent visiting Paris?

This is shown as 165° on the chart.

To find the answer you first have to work out what 165° is as a fraction of 360°.

$\frac{165°}{360°}$ cancels to $\frac{11}{24}$.

To find out how long this is in hours, multiply the fraction by 24 hours giving an answer of 11 hours.

B Calculating angles in a pie chart

This data was collected in a survey of 300 people's favourite holiday destinations:

60 Australia   **40 UK**   **150 Spain**   **50 France**

You can see that Spain is the favourite place. It can also be shown on a pie chart. A pie chart has impact because it is **visual**.

>> **key fact** **To find the angle at the centre of the sector, calculate:** $\frac{\text{number of people}}{\text{total in survey}} \times 360°$

The results are:

Australia $\frac{60}{300} \times 360° = 72°$ UK $\frac{40}{300} \times 360° = 48°$

Spain $\frac{150}{300} \times 360° = 180°$ France $\frac{50}{300} \times 360° = 60°$

C Drawing the pie chart

1 To draw a pie chart, follow these steps:
- Draw a circle.
- Draw a straight line from the centre to the circumference – a radius (in an exam question, one or both of these steps might have been done for you).
- Measure out the angle for the first sector, starting from this line.
- Work round the pie chart one angle at a time.
- Label each sector with its title, and possibly its data value.

2 For the holiday data from section B, the result looks like this (note that there is no particular reason why the destinations should be shown in this order).

The angles have been marked on this diagram, but you would not normally do this.

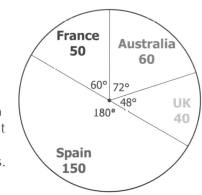

>> practice questions

1 **This table contains data showing the number of films of each category shown in a city:**

classification	18	15	12	PG	U	total
frequency	42	46	56	43	53	

Draw a pie chart to show this information.

2 **Draw a pie chart to show the following data:**

red balloons 30
yellow balloons 50
green balloons 120
white balloons 100

Histograms

When data is in equal class intervals, the heights of the bars are proportional to frequency, and a histogram resembles an ordinary bar chart.

The frequency density of a histogram is calculated as $\dfrac{\text{frequency}}{\text{class width}}$.

A Constructing histograms

>> **key fact** In a histogram, the areas of the rectangles are proportional to the frequencies they represent.

1. There are times when data that is collected forms class intervals with unequal widths.

   This table shows the results of a survey of workers' journey times to work:

Duration of Journey (t minutes)	$0 \leqslant t < 10$	$10 \leqslant t < 15$	$15 \leqslant t < 20$	$20 \leqslant t < 25$	$25 \leqslant t < 30$
Frequency	12	14	20	10	4

2. Now calculate the frequency density of each class interval.
   This is done by using the equation:

   frequency density $= \dfrac{\text{frequency}}{\text{class width}}$.

   In this case, the frequency density represents the number of workers per minute of journey time.

   This means the table now reads:

Duration of Journey (t minutes)	$0 \leqslant t < 10$	$10 \leqslant t < 15$	$15 \leqslant t < 20$	$20 \leqslant t < 25$	$25 \leqslant t < 30$
Frequency density	$12 \div 10 = 1.2$	$14 \div 5 = 2.8$	$20 \div 5 = 4$	$10 \div 5 = 2$	$4 \div 5 = 0.8$

3. These figures give the heights of the bars in the histogram:

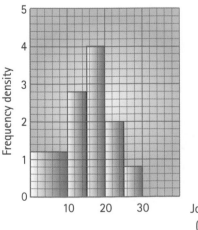

Journey time (t minutes)

B | Interrogating histograms

1 As with all visual ways of displaying data, much useful information can be extracted from a histogram.

>> **key fact** To find the actual frequencies, multiply the frequency density by the class width.

This histogram shows the lifespan of worms in a wormery, before they were sold on:

Lifetime (L weeks)	Frequency density	Frequency
$0 \leqslant L < 4$	16	
$4 \leqslant L < 6$	25	
$6 \leqslant L < 8$	20	
$8 \leqslant L < 10$	10	
$10 \leqslant L < 12$	25	
$12 \leqslant L < 16$	5	

The frequency density can be read straight from the graph.

For example, for the class $0 \leqslant L < 4$, frequency = freq density $\times$ class width = $16 \times 4 = 64$ worms.

2 Calculate the rest of the frequencies for the table above. You should find that the total frequency is 244 worms.

>> practice questions

1 In a survey at Valley College, students were asked how long it had taken them to travel in that morning. This table shows the results of the survey. There are 500 students at the college.

Duration of journey (t minutes)	Frequency
$0 \leqslant t < 5$	55
$5 \leqslant t < 10$	85
$10 \leqslant t < 20$	90
$20 \leqslant t < 30$	150
$30 \leqslant t < 40$	60
$40 \leqslant t < 60$	40

(a) Draw a histogram illustrating this data.

(b) How many students were absent on the day of the survey?

2 Graham grows tomatoes. This histogram shows the heights of the plants in his greenhouses.

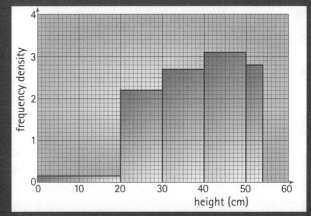

(a) Draw a table to record the classes and frequency density.

(b) Add another column to the table for the frequencies, then calculate them to complete the column.

(c) How many tomato plants does Graham have?

Finding averages

> An average is a single figure that indicates the typical value of a set of data.

> The three types of average are mode, median and mean.

> The procedures for calculating averages are slightly different depending on whether the data are grouped or not.

A The mode

>> **key fact** The mode of a set of data is its most common value.

1 These are the shoe sizes of people in a family: 2, 10, 9, 10, 7, 8, 3, 9, 3, 3.
The mode is 3 because there are three 3s: more than any other number.

2 For data presented in a frequency table, the mode is easy to identify. This table shows the results of a survey on the number of brothers and sisters people have.

Brothers and sisters	0	1	2	3	4	5
frequency	56	138	37	8	0	1

The mode is obviously 1!

3 For data that has been put into groups or 'classes', it is only possible to identify the group with the largest frequency – the modal group or modal class. This frequency table shows the number of pencils found in the pencil cases of a set of students:

Number of pencils	1–9	10–14	15–19	20–24
frequency	13	4	8	2

The modal group is 1–9 pencils.

B The median

>> **key fact** The median of a set of data is its middle value.
The data must be in order of size.

1 Here are the shoe sizes written in order: 2, 3, 3, 3, 7, 8, 9, 9, 10, 10.

There are ten numbers, so there isn't actually a middle one. In this case, use the two middle values and find the number halfway between them: they are 7 and 8, so the median is 7.5.

2 The median for data in a frequency table is easy to find. Using the data on brothers and sisters, totalling the frequencies shows that 250 people were surveyed. The median is halfway between the 125th and 126th data values. The value 1 represents the 59th to 204th data values, so the 125th and 126th are both 1, and so the median is 1.

3 For grouped data, it is possible to find the group that contains the median. For the pencil case data above, the total frequency is 27, so the median is the 14th value. This is in the 10–14 group. To estimate the median more accurately than this for grouped data, see 'Cumulative Frequency' on page 98.

C   The mean

>> **key fact** **The mean of a set of data is the total of the data values divided by the number of data values.**

1 The total of the shoe sizes is 61, so the mean is 61 ÷ 10 = 6.1.

2 For data in a frequency table, the values need to be multiplied by the frequencies, and these results totalled. It is useful to organise these calculations in a table. Here is the table for the brothers and sisters survey:

Brothers and sisters (*x*)	0	1	2	3	4	5	Totals
Frequency (*f*)	56	138	37	8	0	1	250
f × *x*	0	138	74	24	0	5	241

The mean is 241 ÷ 250 = 0.964.

3 **key fact** **For grouped data, the mean can only be estimated. This is because the individual data values are not known.**

To make a sensible estimate, you assume that the data values are evenly spread within each group. This means that the total for each group can be roughly calculated as the middle value or midpoint multiplied by the frequency. Here is the table for the pencil case data:

Number of pencils	1–9	10–14	15–19	20–24	Totals
Midpoint (*x*)	5	12	17	17	22
Frequency (*f*)	13	4	8	2	27
f × *x*	65	48	136	44	293

The estimated mean is 293 ÷ 27 = 10.851… = 10.9 (1 dp).

D   The range

>> **key fact** **The range of a set of data is not an average, but measures how 'spread out' the data values are. It is the difference between the maximum and minimum values.**

The ranges of the data sets used in this section are as follows.

Shoe sizes: 10 − 2 = 8   Brothers and sisters: 5 − 0 = 5
Pencils: 24 − 1 = 23 (this is an estimate because we don't have the exact data)

>> practice questions

Find the mode, median, mean and range of the following data sets.

1 Harry's scores in eight spelling tests: 17, 12, 13, 15, 10, 12, 17, 12.

2 100 bags of crisps were weighed. These are the weights:

3 These are the amounts of money spent by customers in a shop:

Weight (grams)	23	24	25	26	27
Frequency	4	20	35	34	7

Amount spent, £*x*	0 ⩽ *x* < 5	5 ⩽ *x* < 10	10 ⩽ *x* < 15	15 ⩽ *x* < 20	20 ⩽ *x* < 40
Frequency	6	19	21	10	4

Cumulative frequency

Cumulative frequency is a 'running total' of the number of data items below a given value.

The median and quartiles divide a data set into quarters. A box-and-whisker diagram illustrates these statistics clearly.

A Cumulative frequency

>> **key fact** Cumulative frequency is a sort of 'running total' of frequencies. The cumulative frequencies build up as more of the data values are included.

 This table shows the distance, x miles, that 120 people attending a business conference had to travel.

Distance, x miles	$0 < x \leqslant 10$	$10 < x \leqslant 20$	$20 < x \leqslant 30$	$30 < x \leqslant 50$	$50 < x \leqslant 100$	$100 < x \leqslant 140$
Number of people	16	18	37	33	12	4

 The cumulative frequency table looks like this:

Distance, x miles	$x \leqslant 10$	$x \leqslant 20$	$x \leqslant 30$	$x \leqslant 50$	$x \leqslant 100$	$x \leqslant 140$
Number of people	16	34	71	104	116	120

The entry for $x \leqslant 20$ contains the total from the first two groups, the entry for $x \leqslant 30$ the totals for the first three groups, and so on. The final entry is always equal to the total number of data values.

B Cumulative frequency graphs

 These graphs are just plots of the cumulative frequency against the **upper bound** of each group (very important – **not** the midpoint). In the case of the data given above, you would plot the points (10, 16), (20, 34), etc., up to (140, 120).

This is the graph for the distance data:

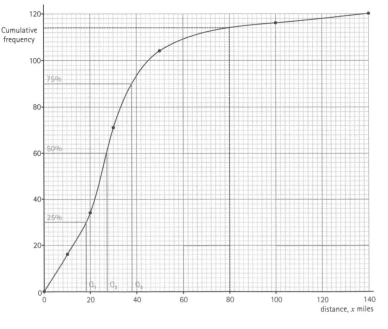

C The median and quartiles

 The **median** of a set of data splits it into two halves.

>> **key fact** To estimate the median, always read from the cumulative frequency value that is 50% of the total.

Follow the green arrow from 60 to Q_2. From this graph, the estimated median distance is 27 miles.

 The **quartiles** split the data into four quarters. The **lower quartile** (Q_1) has 25% of the data below it. The upper quartile (Q_2) has 75% of the data below it. The median is also sometimes called Q_2.

For this data set, follow the green arrows from 30 on the cumulative frequency axis to estimate the lower quartile ($Q_1 = 18$) and from 90 to estimate the lower quartile ($Q_3 = 38$).

>> **key fact** To estimate the quartiles, always read from the cumulative frequency values that are 25% and 75% of the total.

 The **interquartile range** is the range of the central 50% of the data. For the distance data it is $38 - 18 = 20$ miles.

>> **key fact** Interquartile range = $Q_3 - Q_1$.

D Box-and-whisker diagrams

A **box-and-whisker** diagram shows the median, quartiles and the minimum and maximum values of a data set. The 'box' covers the interquartile range and the 'whiskers' the rest of the range. Here is the diagram for the distance data.

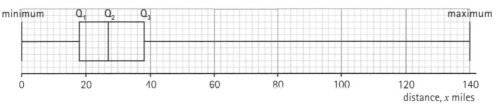

>> practice questions

The following table shows the ages of people living in a village:

Age, A years	Number of people
$0 \leqslant A < 10$	8
$10 \leqslant A < 20$	9
$20 \leqslant A < 30$	11
$30 \leqslant A < 40$	7
$40 \leqslant A < 50$	15
$50 \leqslant A < 60$	22
$60 \leqslant A < 80$	14
$80 \leqslant A < 100$	2

1 Produce a table of the cumulative frequencies for the data.

2 Draw a cumulative frequency graph.

3 Estimate the median, quartiles and interquartile range.

4 Draw a box-and-whisker diagram.

5 (a) Estimate the number of people in the village who are aged 70 or over.

   (b) Estimate the probability that a person selected at random from the village is aged 16 or under.

Comparing sets of data

 Compare sets of data using a measure of average and the range of the data.

 Statistical graphs can help you compare sets of data, by making the average and spread easy to see.

A   Average and spread

key fact   You can compare two sets of data by:
- **checking which is bigger on average (using mode, median or mean)**
- **checking which is more consistent (less spread out, using the range or interquartile range).**

This table shows the last five javelin throws by two athletes, in metres.

Mel	21	26	23	25	26
Kim	18	22	29	24	22

This table compares their results.

	furthest	shortest	mean	range
Mel	26	21	24.2	5
Kim	29	18	23	11

Kim threw the furthest, but also had the shortest throw.

Mel's mean throw was over a metre further than Kim's, and her range is smaller, so her performance was more consistent. If you were choosing a member for the javelin team, you might gamble on Kim, as she had the longest throw, but Mel is probably the better choice as she is more reliable.

B   Back-to-back stem-and-leaf diagrams

This back-to-back stem-and-leaf diagram shows the lifetimes of two types of battery.

```
    Duraplus        XtraPower

           7 |  4 |
           4 |  5 | 4 6
         8 6 2 |  6 | 0 3 8
 9 8 8 6 5 4 4 1 |  7 | 2 2 3 5 7 7 8
     7 7 5 5 2 2 |  8 | 0 0 1 3 3 3 6 9
         8 4 3 |  9 | 5 6 6
           3 2 | 10 |
```

Key: 1 | 0 = 10 hours

Which are the better batteries? Duraplus have the longest lifetime, but their modal stem is only 70. XtraPower's modal stem is 80, so they last longer on average, and their data only covers 5 stems, as opposed to Duraplus's 7, so they are more reliable.

C Comparing graphs

1 This example uses a cumulative frequency graph to compare the time two groups of students spent revising for exams.

Reading from the graph, you can estimate the following statistics:

	median	interquartile range
Group A	1.3	1.8 – 0.7 = 1.1
Group B	1.7	2.6 – 1.0 = 1.6

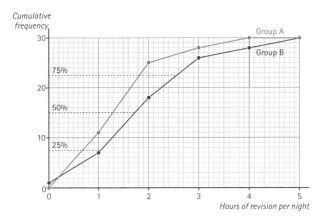

2 Student group A were more consistent in their revision, as the interquartile range is smaller. Group B's median is higher, so they spent longer revising on average.

3 There are many other comparisons you can make. For example, there was a student in Group A who did no revision, but every student in Group B did at least one hour.

>> practice questions

1 Two types of marker pen are tested to see what length of line they write before running out of ink. Ten pens of each type were tested. These are the lengths, in metres, rounded to the nearest 10 m.

Markit	1710	1730	1650	1730	1680	1730	1680	1700	1670	1720
Skribbla	1640	1930	1870	1790	1800	1640	1720	1690	1720	1700

(a) Draw a table, like the one in section A, to compare the results.

(b) If you were going to buy a marker pen, which type would you buy, and why?

2 The same kind of plants were grown using two different fertilisers, Gromore and Sproutwell. The heights the plants grew to are shown in this bar chart, rounded to the nearest centimetre.

(a) How many plants were used in the experiment?

(b) Compare the results of the two fertilisers. Which fertiliser do you think was more effective?

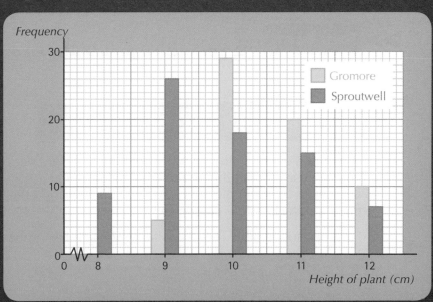

Probability

> A chance event has a number of possible outcomes. Each outcome has a probability, a number between 0 and 1 describing the chance it will happen.

> Probabilities can be expressed as fractions, decimals or percentages.

A Theoretical probability

1 The event 'roll a die' has 6 outcomes, all of which are equally likely – that is, they have the same probability, if the die is fair. This is written

$$P(1) = \frac{1}{6}, P(2) = \frac{1}{6}, \text{ etc.}$$

2 You can use the probability to estimate what will happen when a number of **trials** or experiments are carried out.

>> **key fact** The expected **frequency for an outcome is given by probability × number of trials.**

If you rolled the die 90 times, you would expect each number to come up roughly $\frac{1}{6} \times 90 = 15$ times. This is not a precise prediction, but the more trials you do, the better the prediction is likely to be.

3 You can also predict probabilities when the outcomes are not equally likely. The probabilities for this spinner are

$$P(\text{red}) = \frac{1}{4} \text{ and } P(\text{yellow}) = \frac{3}{4}.$$

B Experimental probability

1 In some circumstances, it isn't possible to find probabilities just by analysing an event mathematically. You need to carry out a number of trials and use the results to estimate probabilities. This gives you the **experimental probability** (sometimes called **relative frequency**).

2 Suppose that in a survey of 500 cars, 35 had at least one defective tyre. You could estimate that the probability that a car checked at random has a defective tyre is $\frac{35}{500} = 0.07$.

The more data you have, the better the estimate of probability is likely to be.

>> **key fact** **Experimental probability for an outcome** $= \dfrac{\text{frequency of outcome}}{\text{number of trials}}$

C Possibility spaces

1 When two events with equally likely outcomes occur together, you can use a **possibility space table** to list all the possible outcomes. For example, if you roll two dice and add the scores, these are the possible outcomes:

Each cell in the table is equally likely, so each has probability $\frac{1}{36}$.

This means that, as 7 occurs 6 times in the table, $P(7) = \frac{6}{36} = \frac{1}{6}$.

		1st die					
		1	2	3	4	5	6
2nd die	1	2	3	4	5	6	7
	2	3	4	5	6	7	8
	3	4	5	6	7	8	9
	4	5	6	7	8	9	10
	5	6	7	8	9	10	11
	6	7	8	9	10	11	12

D   Tree diagrams

1   Where two (or more) events occur together and the probabilities are *not* equal, a tree diagram is useful. Each event has a set of 'branches'. The outcomes are written at the end of each branch, with the probabilities along the branches.

2   This diagram illustrates the following situation: Sunita likes milk chocolates but not dark ones. She takes two sweets at random from a bag containing 6 milk and 4 dark chocolates. What is the probability she will get at least one sweet she likes? Notice that the fractions on the second set of branches have denominator 9. This is because the number of sweets in the bag is different after one has been taken out.

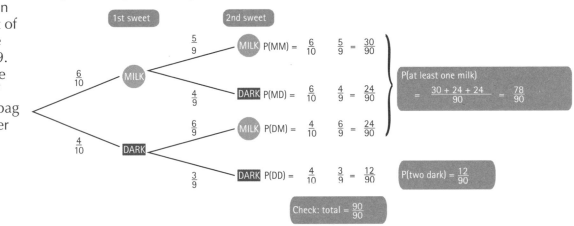

1st sweet   2nd sweet

$\frac{6}{10}$ MILK

$\frac{5}{9}$ MILK $P(MM) = \frac{6}{10} \; \frac{5}{9} = \frac{30}{90}$

$\frac{4}{9}$ DARK $P(MD) = \frac{6}{10} \; \frac{4}{9} = \frac{24}{90}$

$\frac{4}{10}$ DARK

$\frac{6}{9}$ MILK $P(DM) = \frac{4}{10} \; \frac{6}{9} = \frac{24}{90}$

$\frac{3}{9}$ DARK $P(DD) = \frac{4}{10} \; \frac{3}{9} = \frac{12}{90}$

P(at least one milk)
$= \frac{30 + 24 + 24}{90} = \frac{78}{90}$

P(two dark) $= \frac{12}{90}$

Check: total $= \frac{90}{90}$

>> key fact   **To obtain the correct probability for any route through the tree, multiply the probabilities on any branches you use. If the probabilities from two routes have to be combined, add them.**

You could also get the answer by subtracting *P*(two dark) from 1.

---

1   **In a traffic survey, the number of people in each car was counted. These are the results.**

Number of people	1	2	3	4	5
Frequency	144	75	36	33	12

Use this information to estimate the probability that a car picked at random:

(a) contains just the driver;

(b) contains at least three people.

2   **In a charity game, you spin these two spinners. You win a prize if both spinners land on the same colour. Draw a tree diagram to find the probability of winning a prize.**

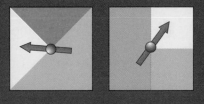

3   **In a game, you roll two dice but score the difference between the numbers showing on the dice. For example, if you roll a six and a two, your score is 6 − 2 = 4.**

(a) Draw a possibility space table to show all the possible scores.

(b) Which score is most likely, and what is its probability?

(c) If the dice are rolled 100 times, how many times would you expect to score 2?

Scatter diagrams and correlation

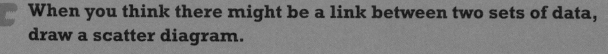

- When you think there might be a link between two sets of data, draw a scatter diagram.

- If there is a link, the points will seem to be scattered close to a line, the line of best fit.

- The line of best fit can be used to make predictions about similar data.

A Scatter diagrams

When it is thought that two sets of data may be linked, a scatter diagram can reveal this.

For example, this table shows the scores obtained by 18 students in two maths tests.

test 1	5	10	15	20	25	30	35	40	45	50	55	60	75	80	85	90	95	100
test 2	4	1	5	16	28	37	46	39	37	57	60	61	86	90	82	81	97	100

The pairs of data values are plotted as co-ordinates.

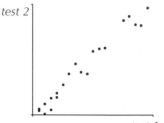

Each data point represents one person's scores. The correlation is positive.

>> **key fact** You should never attempt to join scattered points to each other with straight lines!

B The line of best fit

When your scatter diagram shows that there is a correlation, a **line of best fit** can be added to it.

>> **key fact** There should be roughly equal numbers of points either side of the line. Any points that lie a long way from the line are called outliers.

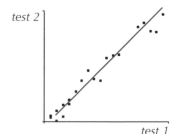

The line of best fit shows the 'trend' of the points. The points are close to the line, so the correlation is strong.

You can use the line of best fit to predict values that follow the same trend. For example, you could answer the question, 'What would someone who scored 70 on test 1 expect to score on test 2?'

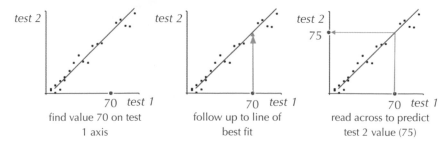

find value 70 on test 1 axis

follow up to line of best fit

read across to predict test 2 value (75)

Warning! Do not use the line of best fit to predict values outside the range of the data – the correlation may not hold for those values.

Different types of correlation

>> **key fact** **Where the points are scattered closely around the line, there is a strong correlation.**

1 If the points are more loosely scattered around the line, there is a moderate correlation. This means that any predictions would be rough estimates, especially if the points are widely scattered, as they are in this diagram.

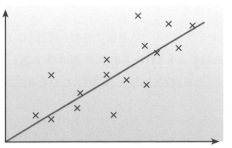

positive correlation

>> **key fact** **If the points are so scattered that there is no obvious line, then there is no correlation.**

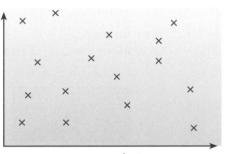

no correlation

2 If the points are scattered from top left to bottom right, the correlation is **negative**. This occurs when, as one quantity increases, the other quantity decreases.

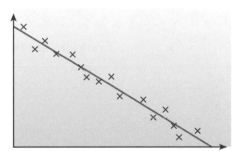

negative correlation

>> practice questions

This table shows the heights and weights of 10 people:

height (cm)	150	152	154	158	159	160	165	170	175	181
weight (kg)	57	64	66	66	62	67	75	69	72	76

1 **Draw the scatter diagram to represent the data. Add a line of best fit.**

2 **Comment on the correlation between height and weight.**

3 **Nisha weighs 58 kg. Estimate her height.**

   (On one axis, use 2 cm to represent 5 cm of height and start at 145 cm. On the other axis, use 2 cm for 5 kg, starting at 55 kg.)

Exam questions and model answers

There are many specific number questions on both papers, however, many of the facts and skills you learn are used in other branches of mathematics. The questions in this section test 'pure' number skills.

>> Specimen question 1

a)   **i)** Show that $\sqrt{5} \times \sqrt{15} = 5\sqrt{3}$. **1 mark**

      ii) Expand and simplify $(\sqrt{5} + \sqrt{15})^2$. **3 marks**

b)   Is ABCD a rectangle?
Show your working clearly. **3 marks**

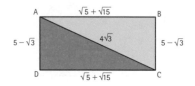

>> Model answer 1

a)   **i)** $\sqrt{5} \times \sqrt{15} = \sqrt{5} \times \sqrt{5} \times \sqrt{3}$ *Use $15 = 5 \times 3$ within the square root*

      $= 5\sqrt{3}$ *Combine the two $\sqrt{5}$ to make 5:* **1 mark**

> •   You could also have done the question this way: $\sqrt{5 \times 15} = \sqrt{75} = \sqrt{25 \times 3} = 5\sqrt{3}$.

a)   **ii)** $(\sqrt{5} + \sqrt{15})^2 = \sqrt{5^2} + \sqrt{15^2} + 2\sqrt{5}\sqrt{15}$ *Use expansion of $(a + b)^2$:* **2 marks**

      $= 5 + 15 + 2 \times (5\sqrt{3})$

      $= 20 + 10\sqrt{3}$ *Simplify correctly:* **1 mark**

> •   This part uses the algebraic identity $(a + b)^2 = a^2 + 2ab + b^2$.

b)   For it to be a rectangle, angle ABC must be a right angle, so by Pythagoras' Rule, $AB^2 + BC^2 = AC^2$.

      $AC^2 = (4\sqrt{3})^2 = 16 \times 3 = 48$.   ***Expand AC^2 correctly:* 1 mark**

      $AB^2 + BC^2 = (\sqrt{5} + \sqrt{15})^2 + (5 - \sqrt{3})^2$

      $= (20 + 10\sqrt{3}) + (25 + 3 - 2 \times 5\sqrt{3})$ *Use the results from (a):* **1 mark**

      $= (20 + 10\sqrt{3}) + (28 - 10\sqrt{3}) = 48$.   ***Simplify $AB^2 + BC^2$ correctly:* 1 mark**

      So $AB^2 + BC^2 = AC^2$, and ABCD is a rectangle.

> •   The key to this part is realising that, although it's a number question about surds, you need a geometrical technique (Pythagoras' Rule) to get you started.

>> Specimen question 2

On a TV quiz show, teams buzz to answer a 'starter' question worth 10 points.

- If they **interrupt** the quizmaster and give a **wrong** answer, they lose 5 points.

- If they get the starter **right**, they also answer **extra** questions worth 5 points each.

- If they get a starter question **wrong**, but **don't interrupt**, they score no points.

  These were the details of one programme.

Who won the game? **3 marks**

	New Bridge College	University of Lakeland
starters correct	16	18
starters wrong	13	9
starters wrongly interrupted	5	11
extra questions	25	26

>> Model answer 2

New Bridge College:

$(16 \times 10) + (13 \times 0) - (5 \times 5) + (25 \times 5) = 160 + 0 - 25 + 125 = 260$ points. **1 mark**

University of Lakeland:

$(18 \times 10) + (9 \times 0) - (11 \times 5) + (26 \times 5) = 180 + 0 - 55 + 130 = 255$ points. **1 mark**

New Bridge College won. *Answer the question:* **1 mark**

- To do this question, calculate the total score for each team.

- This is a Paper 3 question, so the calculations need to be done mentally, or with extra working on the paper. The numbers have been kept easy to help you!

- An alternative (but unorthodox) method is to work out the difference between the teams for each column in the table. This would give Lakeland the following number of points more than New Bridge:
  $(2 \times 10) + (-4 \times 0) - (6 \times 5) + (1 \times 5) = 20 + 0 - 30 + 5 = 25$ (so Lakeland lost).

>> Further questions

1 **a)** Use your calculator to find $\sqrt{7 - 1.5^2}$.

   Write down all the figures on your calculator.

   b) Write your answer to 4 significant figures.

2 Which of the following fractions is nearest to $\frac{3}{4}$?

   $\frac{11}{15}$ $\frac{23}{30}$ $\frac{34}{45}$ $\frac{43}{60}$

3 **(a)** An average human hair is 0.000 05 mm in diameter. Write this number in standard index form.

   (b) One of the hairs from Michelle's head is 48 cm long. Assuming it is of average diameter, how many times longer is it than it is wide?
   Give your answer in standard index form.

4 What fraction is equivalent to the recurring decimal 0.135 353 535...?

>> Answers to further questions

1 (a) 1.903943276...   (b) 1.904

2 $\frac{3}{4} = \frac{135}{180} = 0.75$

$\frac{11}{15} = \frac{132}{180} = 0.7333...$

$\frac{23}{30} = \frac{138}{180} = 0.7666...$

$\frac{34}{45} = \frac{136}{180} = 0.7555...$

$\frac{43}{60} = \frac{129}{180} = 0.7166...$

So $\frac{34}{45}$ is closest.

3 (a) 5×10^{-5} mm

(b) $480 \div (5 \times 10^{-5}) = 9.6 \times 10^6$ times

4 $\frac{67}{495}$

Algebra

Although algebra includes work on co-ordinates and graphs, the main skill you require is to be able to manipulate and transform algebraic expressions. With this skill, you can solve equations, transform formulae and analyse sequences.

>> Specimen question 1

y is inversely proportional to x^2.

(a) When $x = 5$, $y = 30$.

Find the value of y when $x = 10$. **3 marks**

Also, x is proportional to the cube root of u.

(b) When $y = 120$, $u = 0.1$.

Find the value of u when $x = 5$. **4 marks**

>> Model answer 1

(a) $y \propto \frac{1}{x^2}$, so $y = \frac{k}{x^2}$ *Write down the equation of proportionality:* **1 mark**

So $30 = \frac{k}{5^2} = \frac{k}{25}$ *Substitute values*

$k = 750$ *Calculate the constant of proportionality:* **1 mark**

When $x = 10$, $y = \frac{750}{x^2}$

$= \frac{750}{100^2}$ *Substitute for x*

$= 7.5$. *Calculate y:* **1 mark**

- It's important to follow the initial steps of this answer through very carefully. Don't miss out any steps. Check that your equation matches the statement at the beginning of the question.

(b) If $y = 120$, $120 = \frac{750}{x^2}$ *Equation of proportionality*

$120x^2 = 750$ *Rearrange*

$x^2 = \frac{750}{120} = 6.25$

$x = \sqrt{6.25} = 2.5$ *Solve for x:* **1 mark**

$x \propto \sqrt[3]{u}$, so $u \propto x^3$ *Make u the subject*

So $u = kx^3$ *Write down the equation of proportionality:* **1 mark**

So $0.1 = k \times 2.53 = 15.625k$ *Substitute values*

$k = \frac{0.1}{15.625} = 0.0064$ *Calculate the constant of proportionality:* **1 mark**

So $u = 0.0064x^3$

When $x = 5$, $u = 0.0064x$ *Equation of proportionality*

$= 0.0064 \times 5^3$ *Substitute for x*

$= 0.8.$ *Calculate u:* **1 mark**

- The first step involves 'turning round' the proportionality equation from the first part and solving for x.

- Making u the subject of the new proportionality equation isn't absolutely necessary, but can save you a lot of time and difficult manipulation involving cube roots. As the objective is to calculate a value of u, it makes sense to have $u = kx^3$ rather than $x = k\sqrt[3]{u}$.

>> Specimen question 2

(a)   i) Factorise the expression $x^2 - 3x - 18$ **2 marks**

ii) Hence solve the equation $x - 3x = 18$ **1 mark**

(b)   Solve the inequality $x^2 < 25$ **2 marks**

>> Model answer 2

(a)   i) $x^2 - 3x - 18 = (x + ?)(x - ?)$ *The two ?s multiply together to make −18: hence the opposite signs:* **1 mark**

Possibilities are:

$(x + 1)(x - 18)$ *This will give −17x*

$(x + 2)(x - 9)$ *This will give −7x*

$(x + 3)(x - 6)$ *This will give −3x: no need to look any further*

$(x + 6)(x - 3)$

$(x + 9)(x - 2)$

$(x + 18)(x - 1)$

So $x^2 - 3x - 18 = (x + 3)(x - 6)$ *The complete factorisation:* **1 mark**

- When factorising quadratic expressions, always check that you have the correct sign in each bracket.

- Be sure to make a complete list of the factors of the number term (in this case, −18).

---

ii) $x^2 - 3x = 18$ can be rewritten as:

$x^2 - 3x - 18 = 0$ *Write down the equation*

So $(x + 3)(x - 6) = 0$ *From part (i)*

This is only possible if

$(x + 3) = 0$ or $(x - 6) = 0$

So $x = -3$ or $x = 6$ *The complete solution has two answers:* **1 mark**

- If the original expression is equal to 0, its factorised version must be too.

- Quadratic equations that can be factorised like this have two solutions.
  Make sure you give them both!

(b)   The equation $x^2 = 25$ has two solutions,

$x = 5$ and $x = -5$. *Find the two solutions of the equation:* **1 mark**

So $x^2 < 25$ means that $x < 5$, and $x > -5$.

This can be written as a single inequality, $-5 < x < 5$. *Two separate inequalities would do, but this is neater:* **1 mark**

>> Specimen question 3

Alisha and James are investigating hexagonal patterns. Here are the first four.

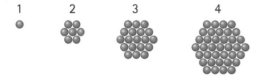

1   2   3   4

They both find a way of generating the numbers of dots.

pattern number	number of dots	Alisha's formula	James's formula
1	1	$3 \times 0 + 1$	$3 \times 1 - 2$
2	7	$6 \times 1 + 1$	$3 \times 4 - 5$
3	19	$9 \times 2 + 1$	$3 \times 9 - 8$
4	37	$12 \times 3 + 1$	$3 \times 16 - 11$
5			
n			

(a)   Fill in the blank cells on row 5. **3 marks**

(b)   Fill in the blank cells on row n. **3 marks**

(c)   Show algebraically that Alisha's and James's formulae are really the same. **2 marks**

>> Model answer 3

(a)   First fill in Alisha's formula. Look at the sequences in the numbers. The first number is 3, 6, 9, 12, **15**. The second number is 0, 1, 2, 3, **4**. The third number is always **1**. So the next entry must be $15 \times 4 + 1$.

Write this down. Don't simplify anything: follow the same pattern as the other entries in the column: **1 mark**

Next fill in James's formula. The first number is always **3**. The second number is 1, 4, 9, 16, **25** (square numbers). The third number is 2, 5, 8, 11, **14**. So the next entry must be $3 \times 25 - 14$.

Write this down. Don't simplify anything: follow the same pattern as the other entries in the column: **1 mark**

Now use either of the formulae to calculate the number of dots. Either one gives 61.

> **Write down this answer: 1 mark**

(b) For Alisha's formula, the first number is three times the pattern number, so this must be $3n$. The second number is one less than the pattern number, so it must be $(n - 1)$. The completed formula is therefore $3n(n - 1) + 1$.

> **Write down this answer: 2 marks. Don't forget the +1 at the end – you'd lose a mark for each missing or incorrect part.**

For James's formula, the second number is the square of the pattern number, so this must be n^2. The third number increases by 3 each time and starts on 2, so it must be $(3n - 1)$. The completed formula is therefore $3n^2 - (3n - 1)$.

> **Write down this answer: 2 marks. Don't forget the 3 at the beginning – you'd lose a mark for each missing or incorrect part. The formula could also be written.**

(c) To show that the formulae are identical, expand and remove all brackets.

For Alisha's formula, $3n(n - 1) + 1 = 3n^2 - 3n + 1$. **Correct expansion: 1 mark**

For James's formula, $3n^2 - (3n - 1) = 3n^2 - 3n + 1$. These are the same. **Correct expansion: 1 mark**

- Successfully answering this question hinges on being able to identify the various sequences that have been used to make up the formulae, then writing them correctly using algebra. Although the final result (a three-term quadratic formula) is quite complex, the individual parts are easy to construct.

>> Further questions

1 Solve the equation $\frac{4x + 3}{2} - \frac{x - 10}{5} = 26$

2 Solve the simultaneous equations

$3x + 2y = 4$

$y^2 - 2x^2 = 17$

3 Make y the subject of the formula $x = \frac{y - 4}{1 + y}$.

4 Solve the equation $x^3 - 4x = 10$ by trial and improvement, correct to 1 decimal place.

>> Answers to further questions

1 $x = 12.5$

2 $x = -2, y = 5$ or $x = 26, y = -37$.

3 $y = \frac{x + 4}{1 - x}$.

4

x	$x^3 - 4x$	comments
1	-3	$x > 1$
2	0	$x > 2$
3	15	$x < 3$
2.5	5.625	$x > 2.5$
2.6	7.176	$x > 2.6$
2.7	8.883	$x > 2.7$
2.8	10.752	$x < 2.8$
2.75	9.796875	$x > 2.75$

$x = 2.8$ to 1 dp

Shape, space and measures

This part of mathematics covers many different topics, including all work on angles, areas and volumes of shapes, and anything to do with measurement, including speed.

>> Specimen question 1

ABC is a triangle with a point O located inside it.

$\overrightarrow{OA} = \mathbf{a}$, $\overrightarrow{OB} = \mathbf{b}$, $\overrightarrow{OC} = \mathbf{c}$.

The midpoints of AB, BC and CA are J, K, and L, respectively.

(a) Find, in terms of **a**, **b** and **c**, the vectors representing:

    i) $\overrightarrow{AC}$ **1 mark**

    ii) $\overrightarrow{JK}$ **4 marks**

(b) Write down, in terms of **a**, **b** and **c**, the vectors representing $\overrightarrow{KL}$ and $\overrightarrow{LJ}$ **3 marks**

(c) i) Describe the relationship between triangle ABC and triangle JKL. Give reasons for your answer. **2 marks**

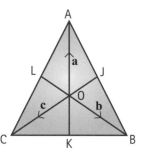

    ii) The area of triangle JKL is 8.5 cm². What is the area of triangle ABC? **1 mark**

>> Model answer 1

(a) i) $\overrightarrow{AC} = \mathbf{c} - \mathbf{a}$ $(-\overrightarrow{OA} + \overrightarrow{OC})$: **1 mark**

- This is a fact you should memorise.

    ii) $\overrightarrow{JK} = \overrightarrow{OK} - \overrightarrow{OJ}$ **1 mark**

$$= \tfrac{1}{2}(\mathbf{b} + \mathbf{c}) - \tfrac{1}{2}(\mathbf{a} + \mathbf{b})$$

$$= \tfrac{1}{2}\mathbf{b} + \tfrac{1}{2}\mathbf{c} - \tfrac{1}{2}\mathbf{a} + \tfrac{1}{2}\mathbf{b} \quad \textit{Correct procedure for midpoints: } \textbf{1 mark}$$

$$= \tfrac{1}{2}\mathbf{c} - \tfrac{1}{2}\mathbf{a}$$

$$= \tfrac{1}{2}(\mathbf{c} - \mathbf{a}) \quad \textit{Simplify correctly: } \textbf{1 mark}$$

- The location of a midpoint is also a fact you should memorise.

(b) $\overrightarrow{KL} = \tfrac{1}{2}(\mathbf{a} - \mathbf{b})$ *Realise that this is like the last part:* **2 marks**

- You can do this the 'long' way, but it's much easier just to realise that the procedure is the same as that in **(a(ii))**, using two other points.

    $\overrightarrow{LJ} = \tfrac{1}{2}(\mathbf{b} - \mathbf{c})$ **1 mark**

(c) **i)** The vectors for the sides of JKL are half those of the corresponding sides in ABC. JKL is therefore similar to ABC and a quarter of its area. **1 mark for each fact**

> • You need to spot that $\overrightarrow{CB}$ and $\overrightarrow{BA}$ are related to the vectors you just found.

ii) $4 \times 8.5\,cm^2 = 34\,cm^2$ *JKL is double the size of ABC, so the area is 4 times larger:* **1 mark**

> • An enlargement with scale factor s increases the area of a shape by a factor of s^2.

>> Specimen question 2

The diagram shows a box in the shape of a cuboid. A metal rod 1 m long has been placed in the box with one end at E and the other resting against edge CG, at P.

(a) How far below C is the top end of the rod?

Give your answer to the nearest centimetre. **5 marks**

(b) Calculate angle PEG, the angle of inclination of the rod to the horizontal.

Give your answer to the nearest degree. **2 marks**

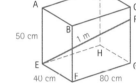

>> Model answer 2

(a) From triangle EFG, by Pythagoras' Rule,
$EG^2 = d^2 = 40^2 + 80^2 = 8000$ *You either write this down, draw a diagram like the one below, or mark these lengths on the original diagram:* **1 mark**; *then calculate d^2:* **1 mark**

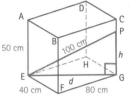

Using Pythagoras' Rule in triangle PEG,
$h^2 = EP^2 - d^2 = 100^2 - 8000 = 2000$ *State Pythagoras' Rule and substitute values:* **1 mark**

$h = \sqrt{2000} = 44.721\ldots$ cm

Required length PC $= 50 - h = 5.278\ldots$ cm

$= 5\,cm$, to nearest cm *Identify that you need Pythagoras:* **2 marks**

> • It's important to identify the two triangles you're going to use.
>
> • In the first triangle, you're going to re-use d^2, so there's no need to calculate d itself.
>
> • Remember that you're not finding the hypotenuse in triangle PEG, so you subtract values in Pythagoras' Rule.

(b) In triangle PEG, tan PEG $=$ opposite $\div$ adjacent *State the correct trig ratio and ...*

$= h \div d$

$= \sqrt{2000} \div \sqrt{8000}$ *Substitute correctly:* **1 mark**

$= 0.5$

So angle PEG   = tan⁻¹(0.5) *Use the inverse trig ratio to calculate an angle*

= 26.565...° *Calculate the angle*

The rod makes an angle of 27° with the horizontal, to the nearest degree. ***Give the rounded answer: 1 mark***

- The crucial thing here, of course, is to use the right trig ratio!
- Remember to keep the values of trig ratios in your calculator memory ready for the inverse calculation, so there are no rounding errors.

>> Further questions

1   A circle of radius 8 cm, centre O, has a regular octagon ABCDEFGH positioned so that its vertices are points on the circumference.

   (a) Calculate the area of the minor sector OGHAB. Give your answer in terms of π.

   (b) Calculate the perimeter of the minor sector OGHAB. Give your answer in terms of π.

   (c) M is the midpoint of AB. By considering triangle OBM, calculate the area of the octagon. Give your answer to the nearest square centimetre.

2   The diagram shows the dimensions of a pentagonal prism.

   (a) Sketch the net of the prism. Mark on relevant measurements.

   (b) Calculate the surface area of the prism.

   (c) Calculate the volume of the prism. Give your answer in cubic metres.

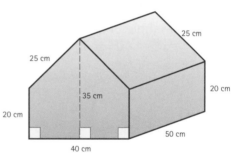

>> Answers

1   **(a)** 24π cm²

   (b) $16 + 6\pi$ cm

   (c) 181 cm²

2   **(a)**

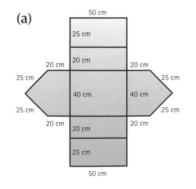

   (b) One side $= (40 \times 20) + \frac{1}{2}(40 \times 15) = 1100$ cm²

   Rectangles $= 130 \times 50 = 6500$ cm²

   Total $= 8700$ cm²

   (c) $1100 \times 50 = 55\,000$ cm³ $= 0.055$ m³

Handling data

Handling data covers three main areas: representing data, which involves displaying data in charts and tables, and interpreting them; processing data, which includes work on averages and ranges to describe and compare frequency distributions; and probability, the study of chance events.

>> Specimen question 1

(a) In a game, these two spinners are spun at the same time.

You multiply the number on one spinner by the number on the other.
Draw a table to list all the possible outcomes. **2 marks**

(b) If you score 4 or less, you lose. If you score 6 or more, you win. If you score 5, you have another go. You are allowed a maximum of three goes. On the third go, scoring 5 or over counts as a win. Draw a tree diagram to show the probabilities for this game. **3 marks**

(c) What is the probability of winning the game? **3 marks**

>> Model answer 1

(a)

		2nd spinner				
		1	**2**	**3**	**4**	**5**
1st spinner	**1**	1	2	3	4	5
	2	2	4	6	8	10

1 mark for each result row: 2 marks

(b)

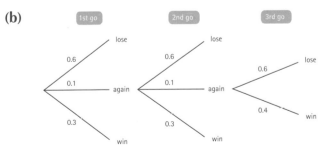

Each 'go' is worth 1 mark: 3 marks

(c) Probability of winning on first go = 0.3.

Probability of winning on second go
= 0.1 × 0.3 = 0.03. ***Multiply probabilities along the branches: 1 mark***

Probability of winning on third go
= 0.1 × 0.1 × 0.4 = 0.004. ***Multiply probabilities along the branches: 1 mark***

Total = 0.334 ***Add the probabilities of the alternatives: 1 mark***

- It's important to set out the table in **(a)** like this, so that the duplicated scores (2 and 4) appear twice each. A simple list wouldn't help you find the probabilities.

- Remember that not every branch on a tree diagram has to continue on to the next event. If you win or lose on a certain go, that branch will terminate. On the exam paper, you would probably be given the structure of the diagram and just have to fill in the probabilities.

- Remember when to multiply probabilities (things happening together) or to add them (one thing or the other happens). Note that you can check the answer by finding the probability of losing:
  0.6 + (0.1 × 0.6) + (0.1 × 0.1 × 0.6) = 0.666.
  P(lose) + P(win) = 0.666 + 0.334 = 1, as you would expect.

>> Further questions

1   A medical treatment has a 0.7 probability of success. If it doesn't work, it can be repeated, but the probability of success is reduced by 0.1 each time. Once the probability of success drops below 50%, doctors are unwilling to prescribe the treatment.

   (a) Draw a tree diagram to illustrate this situation.

   (b) What is the probability that the treatment will eventually succeed?

2   In a game at a charity stall, you pay 20p to spin these two fair spinners:

   (a) Draw a possibility space diagram to show the possible outcomes.

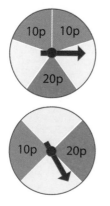

   (b) Copy and complete this table:

win	0p	10p	20p	30p	40p
probability					$\frac{1}{20}$

>> Answers to further questions

1   (a)

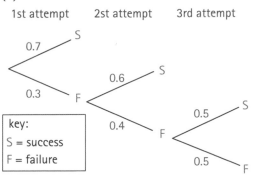

1st attempt   2st attempt   3rd attempt

key:
S = success
F = failure

(b) 0.94

2   (a)

		1st spinner				
		0p	**0p**	**10p**	**10p**	**20p**
2nd spinner	**0p**	0p	0p	10p	10p	20p
	0p	0p	0p	10p	10p	20p
	10p	10p	10p	20p	20p	30p
	20p	20p	20p	30p	30p	40p

(b)

win	0p	10p	20p	30p	40p
probability	$\frac{1}{5}$	$\frac{3}{10}$	$\frac{3}{10}$	$\frac{3}{20}$	$\frac{1}{20}$

Answers to practice questions

Number

The decimal number system [p 3]

1 (a) $25 \times 3 = 75 \to 7.5$ (b) $3 \times 12 = 36 \to 0.36$

  (c) $64 \div 8 = 8 \to 0.8$ (d) $144 \div 3 = 48 \to 48$

2 (a) -7 (b) 2 (c) 19 (d) 4

  (e) -14 (f) 5 (g) -32 (h) 21

  (i) -5 (j) 8 (k) 100 (l) -27

3 (a) -134 (b) -58 (c) -649

  (d) -2.3 (e) -7650 (f) -50

  (g) 0.04 (h) 0.0025

Fraction calculations [p 5]

1 $\frac{1}{2}\left(\frac{60}{120}\right)$, $\frac{8}{15}\left(\frac{64}{120}\right)$, $\frac{7}{12}\left(\frac{70}{120}\right)$, $\frac{3}{5}\left(\frac{72}{120}\right)$, $\frac{5}{8}\left(\frac{75}{120}\right)$

2 (a) $\frac{9}{10}$ (b) $\frac{3}{16}$ (c) $1\frac{11}{20}$ (d) $\frac{2}{3}$

  (e) $5\frac{7}{8}$ (f) $4\frac{7}{9}$ (g) $\frac{2}{15}$ (h) $\frac{5}{12}$

  (i) $1\frac{1}{4}$ (j) $2\frac{1}{2}$ (k) $1\frac{2}{3}$ (l) 8

3 (a) 28 km   (b) 60 ml   (c) £12.10   (d) 1125 g

4 (a) $\frac{1}{3}$ (b) $\frac{2}{3}$ (c) $\frac{3}{5}$ (d) $\frac{21}{50}$

Fractions, decimals and percentages [p 7]

1

	fraction	decimal	percentage
(a)	$\frac{3}{5}$	0.6	(60%)
(b)	$\left(\frac{9}{10}\right)$	0.9	90%
(c)	$\frac{7}{20}$	(0.35)	35%
(d)	$\left(\frac{1}{20}\right)$	0.05	5%
(e)	$\frac{11}{25}$	(0.44)	44%
(f)	$\frac{1}{8}$	0.125	(12.5%)
(g)	$\left(\frac{3}{16}\right)$	0.1875	18.75% $(18\frac{3}{4}\%)$
(h)	$\frac{1}{500}$	0.002	(0.2%)
(i)	$2\frac{3}{8}$	(2.375)	237.5%
(j)	$\frac{9}{125}$	(0.072)	7.2%
(k)	$\left(\frac{1}{9}\right)$	$0.\dot{1}$	11.1% $(11\frac{1}{9}\%)$
(l)	$1\frac{9}{20}$	1.45	(145%)

2 (a) (i) $0.41\dot{6}$ (ii) $0.\dot{2}8571\dot{4}$ (iii) $0.\dot{1}$

  (b) (i) $\frac{4}{9}$ (ii) $\frac{11}{90}$ (iii) $\frac{199}{330}$

Powers and roots [p 9]

1 (a) 512 (b) 484 (c) $15\,625$ (d) $\frac{1}{32}$ or 0.03125

  (e) 8 (f) 2.65 (g) 3 (h) 2.15

2 (a) 5 (b) 10 (c) 5 (d) -3

  (e) 8 (f) 9 (g) 2 (h) 10

3 (a) 3 (b) 5 (c) 2 (d) 10

4 (a) 4 (b) 9 (c) $\frac{1}{2}$ (d) $\frac{1}{343}$

Surds [p 11]

1 $4\sqrt{2}$ **2** $10\sqrt{3}$ **3** $7\sqrt{5}$

4 $\frac{\sqrt{3}}{4}$ **5** $\frac{2\sqrt{2}}{3}$ **6** $\sqrt{\frac{2}{100}} = \frac{\sqrt{2}}{10}$

7 $3\sqrt{7} + \sqrt{21}$ **8** $\sqrt{2} - 2$ **9** $4\sqrt{10} + 30$

10 $8 + 3\sqrt{6}$ **11** $6 - 5\sqrt{3}$ **12** $60 - 52\sqrt{5}$

13 $\frac{3\sqrt{11}}{11}$ **14** $\frac{\sqrt{6}}{15}$ **15** $\frac{1}{\sqrt{3}}$

16 $3 + 2\sqrt{3}$ **17** $-\frac{3 + \sqrt{5}}{2}$

Standard index form [p 13]

1 (a) 2×10^4 (b) 4×10^6 (c) 5.5×10^5
  (d) 9.71×10^{10} (e) 5×10^{-2} (f) 3×10^{-7}

  (g) 7.2×10^{-4} (h) 3.55×10^{-1}

2 (a) $60\,000\,000$ (b) $323\,000$

  (c) $19\,000\,000\,000$ (d) 40

  (e) 0.00007 (f) 0.0000000199

  (g) 0.0903 (h) 0.0000000008

3 (a) 8×10^7 (b) 5×10^1 (c) 4.9×10^4

  (d) 4.89×10^8

4 (a) 2.25×10^{10} (b) 7.01×10^{10}

  (c) -1.004×10^3 (d) 2.5×10^{-4}

Ratio and proportion [p 15]

1 (a) $5 : 1$ (b) $3 : 2$

  (c) $3 : 4$ (d) $5 : 6$

2 (a) $1 : 2.5$, $0.4 : 1$ (b) $1 : 0.6$, $1\frac{2}{3} : 1$

3 (a) £60 : £90   (b) £135 : £15

4 125 calories

5 0.48 m $= 48$ cm

Percentage calculations [p 17]

1

	£200	£11	25 litres	500 tonnes
Increase by 10% …	£220	£12.10	27.5 l	550 t
Increase by 3% …	£206	£11.33	25.75 l	515 t
Decrease by 20% …	£160	£8.80	20 l	400 t
Decrease by 33% …	£134	£7.37	16.75 l	335 t

2 (a) 70%   (b) 15%   (c) 4%   (d) 32%

3 (a) 20% increase   (b) 6% increase

   (c) 26% decrease   (d) 1.5% decrease

4 (a) £35   (b) £36   (c) £81.25

   (d) £3300

Algebra

Algebraic expressions [p 19]

1 (a) 2   (b) 1   (c) 3   (d) 1

   (e) 1   (f) 1   (g) 3   (h) 1

   (i) 1   (j) 4

2 (a) equation (b) identity (c) equation (d) formula

   (e) equation (f) formula (g) identity (h) equation

   (i) formula (j) identity

Formulae and substitution [p 21]

1 (a) 15   (b) 42   (c) 2   (d) 0.01

   (e) 3   (f) 100   (g) –32   (h) 1600

   (i) 40   (j) 6 (or –6)

2 (a) 0.6   (b) 0.2   (c) 0.1875   (d) 3

   (e) 0.3

3 (a) $B = \frac{w}{h^2}$ (b) $S = 2(lw + wh + hl)$

Rearranging formulae [p 23]

1 $x = r - 9$ **2** $x = a + z$

3 $x = \frac{16 - 4y}{5}$ **4** $x = \sqrt{\frac{c}{m}}$

5 $x = 4m$ **6** $x = p^2 + pc$

7 $x = \frac{cb}{m}$ **8** $x = \frac{5 - p}{3 + f}$

9 $y = \frac{9 - 2x}{5}$ **10** $y = \frac{x}{2} - 5$

Using brackets in algebra [p 25]

1 $x = 4$ **2** $x = 10$

3 $x = -8$ **4** $x = 4.5$

5 $x = -5\frac{1}{3}$ **6** $x = 4$

7 $x = 15\frac{1}{2}$

8 (a) $7x(2x + 1)$ (b) $9y(4y - 1)$

9 (a) $5y^2(3y^2 + 5)$ (b) $20a(5a + b^3)$

Factorising quadratic expressions [p 27]

1 $x(x + 5)$ **2** $4x(x - 3)$

3 $(x + 4)(x + 2)$ **4** $(t - 7)(t + 2)$

5 $(5x - 4)(5x + 4)$ **6** $(2x - 3)(x + 4)$

7 $(2h - 3)(2h + 1)$ **8** $(4x + 3)(2x - 5)$

9 $3(x + 4)(x + 1)$ **10** $6(n + 6)(n - 3)$

11 $5(2k + 3)(2k - 3)$ **12** $2(3x - 5)(2x + 7)$

Algebraic fractions [p 29]

1 $\frac{y^3}{z}$ **2** $1\frac{1}{5}$ **3** $\frac{1}{2x - 8}$ **4** $5x^2$

5 $4m^2 x^3$ **6** $\frac{6a + 7}{20}$ **7** $\frac{5t + 8}{6}$

Solving equations [p 31]

1 $x = 7$ **2** $k = 5$

3 $b = 3.6$ **4** $x = 3$

5 $u = 5$ **6** $y = \frac{-13}{4} = -3.25$

7 $N = \frac{5}{4} = 1.25$ **8** $z = 2$

9 $x = 2.5$ (or -2.5)   **10** $x = 32$

Equations of proportionality [p 33]

1 $y = \frac{3x^2}{512}$

speed (x km/h)	41.3 (1 dp)	64	80
braking distance (y metres)	10	24	37.5

2 $v = \frac{5\sqrt{30}}{\sqrt{d}}$

planet	Mercury	Earth	Neptune
orbital distance (d au)	0.33 (2 sf)	1	30
speed (v km/s)	48	27 (2 sf)	5

Trial and improvement [p 35]

1 2.2   **2** 2.1   **3** 3.5, –0.1

4 1.8, –0.8   **5** 1.7, –0.3, –1.4   **6** 2.3

Quadratic equations [p 37]

1 $x = -1, 6$

2 $x = 2 \pm \sqrt{11} = -1.32, 5.32$ (2 dp)

3 $x = -\frac{1}{6} \pm \frac{\sqrt{85}}{6} = -1.70, 1.37$

4 $x = -2.5, 1$

5 $x = \frac{2}{5} \pm \frac{\sqrt{14}}{5} = -0.35, 1.15$ (2 dp)

6 $x = 0, 3.5$

Functions [p 39]

1 **(a)** 5   **(b)** −11   **(c)** −0.8

   (d) 16   **(e)** 0   **(f)** $-\frac{1}{2}$

2 **(a)** $2x + 1$ **(b)** $4x - 1$ **(c)** $2x^2 - 1$

   (d) $(x - 2)^2$ **(e)** $\frac{1}{x^2}$ **(f)** $\frac{1}{x}$

   (g) $\frac{2}{2+x}$ **(h)** $4x - 3$

Inequalities and regions [p 41]

1 **(a)** $x \leqslant 2$

   (b) $x > -3\frac{1}{2}$

   (c) $h > 4$

   (d) $-15 < t \leqslant 9$

   (e) $x \leqslant -6$ or $x \geqslant 6$

   (f) $-5 < x < 4$

   (g) $x \leqslant -2 - \sqrt{5}$ (−4.24, 2 dp)
       or $x \geqslant -2 + \sqrt{5}$ (0.24, 2 dp)

2 The vertices of the triangle are (–3, 1), (1, 5) and (5, 1).

Number patterns and sequences [p 43]

1 **(a)** 13   **(b)** −2   **(c)** 15   **(d)** 80

   (e) 5, 10   **(f)** 37

2 **(a)** 4, 14, 24, 34, 44, …

   (b) 5, 1, −3, −7, −11, …

   (c) 50, 25, 12.5, 6.25, 3.125, …

   (d) 3, 5, 11, 29, 83, …

3 **(a)** 2, 6, 10, 14, …   **(b)** 11, 14, 19, 26, …

   (c) 3, 9, 27, 81, …   **(d)** 2, 8, 20, 40, …

Sequences and formulae [p 45]

1 **(a)** First term 6, add 5

   (b) $u_n = 5n + 1$

   (c) 51

2 **(a)** First term 3, add 6

   (b) $u_n = 6n - 3$

   (c) 57

3 **(a)** First term 10, subtract 1

   (b) $u_n = -n + 11$ or $11 - n$

   (c) 1

4 **(a)** First term 1, add 0.2

   (b) $u_n = 0.8 + 0.2n$

   (c) 2.8

5 **(a)** First term 2, add 3, 4, 5, etc.

   (b) $u_n = \frac{n(n + 3)}{2}$

   (c) 65

6 **(a)** First term 12, multiply by 4

   (b) $u_n = 3 \times 4n$

   (c) 3 145 728

7 **(a)** First term 0, add 3, 5, 7, etc.

   (b) $u_n = n^2 - 1$

   (c) 99

8 **(a)** —

   (b) $u_n = 2n^3$

   (c) 2000

9 **(a)** First two terms 11 and 17, add previous two terms

   (b) —

   (c) 809

10 **(a)** —

   (b) $u_n = \frac{2n - 1}{2^n}$

   (c) $\frac{19}{1024}$

Lines and equations [p 47]

1 **(a)** gradient = 7, y-intercept = 3

   (b) gradient = 3, y-intercept = −5

   (c) gradient = 1, y-intercept = 12

   (d) gradient = 3, y-intercept = −2

2 **(a)** gradient = −3, y-intercept = 4

   (b) gradient = 2, y-intercept = −7

   (c) gradient = $\frac{1}{3}$, y-intercept = $-\frac{7}{3}$

   (d) gradient = $-\frac{1}{5}$, y-intercept = $\frac{9}{5}$

3 $y = 4x + 5$

4 $y = x + 6$

5 $y = 2x + 1$

6 Point of intersection = (3, 2). Equation is $y = \frac{1}{2}x + \frac{1}{2}$.

Curved graphs [p 49]

1 **(a)**

x	-3	-2	-1	0	1	2	3	4	5
y	14	7	2	-1	-2	-1	2	7	14

Solutions: $x = 2.4$ and $x = -0.4$

(b)

x	-3	-2	-1	0	1	2	3
y	17	2	-7	-10	-7	2	17

Solutions: $x = 1.8$ and $x = -1.8$

2 **(a)** add the line $x = 5$; solutions $x = -1.6$ and 3.6.

(b) add the line $y = 5 - 2x$; solutions $x = -2.6$ and 1.9.

Transforming graphs [p 51]

1 **(a)** $y = x^2 + 4$ **(b)** $y = (x + 3)^2 = x^2 + 6x + 9$

(c) $y = x^2 - 5$ **(d)** $y = (x - 7)^2 = x^2 - 14x + 49$

2 **(a)** $y = 3x^3$ **(b)** $y = \frac{1}{2}x^2 = \frac{x^2}{2}$

(c) $y = (\frac{1}{2}x)^2 = \frac{x^2}{4}$ **(d)** $y = (10x + 1)^2$

(e) $y = (\frac{x}{5})^3 + 5(\frac{x}{5})^2 = \frac{x^3}{125} + \frac{x^2}{5}$

3 **(a)** 5× vertical stretch

(b) 3× horizontal stretch

(c) squash to half size horizontally

(d) 2× horizontal stretch

(e) 10× vertical stretch or 10× horizontal stretch!

Linear simultaneous equations [p 53]

1 **(a)** $x = 5, y = 2$ **(b)** $m = 6, n = 1$

(c) $t = 6, r = 6$ **(d)** $x = 5, y = 4$

(e) $c = 7, d = 9$

2 **(a)** $x = 3, y = 7$ **(b)** $x = 2.6, y = -1.2$

Mixed simultaneous equations [p 55]

1 $x = 4, y = 3$ or $x = 3, y = 4$

2 $x = -2.5, y = -2$ or $x = 1, y = 5$

3 $x = 2, y = 2$ or $x = 0, y = 0$

4 $x = 4.93, y = 2.07$

5 $x = 3, y = 4$ or $x = -\frac{1}{2}, y = 7\frac{1}{2}$

Shape, space and measures

Rounding and accuracy [p 57]

1 **(a)** $(30 - 20) \times 2 = 20$:   35.4

(b) $\frac{10 \times 40}{10} = 40$:   34.3

(c) $\frac{4}{2} \times \frac{2}{4} = 1$:   1.11

2 **(a)** 9.5 cm ⩽ length < 10.5 cm;
4.5 cm ⩽ width < 5.5 cm;
42.75 cm² ⩽ area < 57.75 cm²

(b) 15.75 mm ⩽ length < 16.25 mm;
6.25 mm ⩽ width < 6.75 mm;
98.4375 mm² ⩽ area < 109.6875 mm²

3 **(a)** 995 l ⩽ volume < 1005 l,
i.e. 0.995 m³ ⩽ volume < 1.005 m³

(b) 0.75 m² ⩽ area < 0.85 m²

(c) 1.17 m (2 dp) ⩽ depth < 1.34 m

Dimensions [p 59]

1 **(a)** $2\pi r$ length **(b)** πr^2 area

(c) $\pi r^2 h$ volume **(d)** $2\pi r h$ area

2 There are too many dimensions.
For it to be area, it must have only 2 dimensions.

3 It must represent a volume since it is of dimension three.

Speed and motion graphs [p 61]

1

	distance	time	average speed
(a)	(30 km)	(2 hours)	15 km/h
(b)	(4 km)	(30 minutes)	8 km/h
(c)	(100 km)	$2\frac{1}{2}$ hours	(40 km/h)
(d)	12.5 metres	(0.05 seconds)	(250 m/s)
(e)	28 800 km	(1 hour)	(8 km/second)
(f)	(400 m)	(1 minute)	24 km/h

2 **(a)**

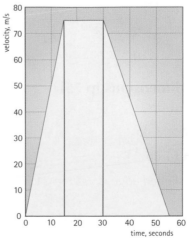

(b) 2625 m

Properties of shapes [p 63]

1 (a) 32° **(b)** 96° **(c)** 71° **(d)** 200°

2 (a) 1800° **(b)** 3240°

3 (a) 140° **(b)** 40° **(c)** 9

Circle geometry [p 65]

1 $a = 61°$

2 $x = 41°, y = 39°$

3 $x = 60°, 2x = 120°$

4 $x = 23°, y = z = 66°$

Pythagoras' Rule and basic trigonometry [p 67]

1 (a) 10.63 cm (2 dp) **(b)** 10.01 cm (2 dp)

  (c) 6.88 cm (2 dp) **(d)** 10.40 cm (2 dp)

2 (a) 62.2° (1 dp) **(b)** 30°

  (c) 47.3° (1 dp) **(d)** 58.4° (1 dp)

The sine rule [pp 69]

1 $c = 10.06$ cm

2 $WY = 13.12$ cm

3 $c = 38.77°$ or $142.23°$

4 $z = 34.25°$ or $145.75°$

5 689.51 cm²

6 43.30 cm²

The cosine rule [p 71]

1 (a) 44.42° **(b)** 18.57°

  (c) 95.22° **(d)** 94.41°

2 (a) 14 **(b)** 10.44

3 118.21m

4 177.4 km

Trigonometric graphs: angles of any size [p 73]

1 (a) 26°, 334°, 386°, 694°

  (b) −131°, −49°, 229°, 311°

  (c) 72°, 252°, 432°

2 (a) 20°, 160°, 380°, 520°

  (b) 100°, 280°

  (c) −272°, −88°, 88°, 272°

3 (a) $\frac{1}{\sqrt{2}}$ **(b)** $\sqrt{3}$ **(c)** $-\frac{\sqrt{3}}{2}$

  (d) $\frac{\sqrt{3}}{2}$ **(e)** $\frac{1}{\sqrt{2}}$ **(f)** $\frac{1}{\sqrt{3}}$

Problems in three dimensions [p 75]

1 (a) $\sqrt{29} = 5.4$ cm to 1 dp

  (b) $\sqrt{525} = 22.9$ cm to 1 dp

  (c) $\sqrt{3} = 1.7$ units to 1 dp

2 The length of each slanted edge is $\sqrt{450} = 21.2$ cm to 1 dp.

  The total length is therefore $40 + 4\sqrt{450} = 124.9$ cm to 1 dp or 1.249 m.

3 $AC = \frac{10}{\tan 16°}$ $BC = \frac{10}{\tan 32°}$

  $\tan \angle CAB = \frac{BC}{AC} = \frac{10}{\tan 32°} \div \frac{10}{\tan 16°} = \frac{\tan 16°}{\tan 32°}$

  So $\angle CAB = \tan^{-1} \frac{\tan 16°}{\tan 32°} = 24.6°$ to 1 dp. So the bearing of B from A is 025°, to the nearest degree.

Calculating areas [p 77]

1 17.5 m²

2 206 m²

3 3 rolls

4 7 tins

5 22.75 cm²

6 10 cm

Circle calculations [p 79]

1 (a) 43.98 cm, 153.94 cm² **(b)** 12.57 cm, 12.57 cm²

  (c) 78.54 cm, 490.87 cm² **(d)** 68.49 cm, 373.25 cm²

2 (a) 15.90 m² **(b)** 21.49 m²

3 5.64 cm

Volume calculations [p 81]

1 60 m³ **2** 108 m³

3 3770 cm³ **4** 0.62 m³

5 78 m³ **6** 49 763 cm³ (49 800 cm³)

Congruence and similarity [p 83]

1 RHS congruence test works

2 12; area scale factor = 4

3 $a = 9.2$ cm, $b = 11.5$ cm; volume scale factor = 12.167

4 Neither. Congruence tests do not apply.

Constructions [p 85]

All answers in this exercise should be self-checking.

Loci [p 87]

1 (a–b)

  RS = 9 cm, RT = 6 cm, ST = 7.5 cm

  Circle radii: R = 5 cm, S = 6 cm, T = 3.5 cm

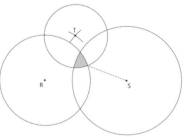

Diagram shown is reduced to fit – use for reference only

(c) 46 km (allow 2 km either way)

2 (a–b)

Triangle sides 7 cm

Locus distance from triangle = 1.5 cm

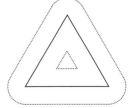

Diagram shown is reduced to fit
– use for reference only

3 (a) $y = 5$ [passes through (0, 5) and (10, 5)]

(b) $y = 10 - x$ [passes through (0, 10) and (10, 0)]
(or $x + y = 10$)

(c) (5, 5)

Vectors [p 89]

1 (a) a + **b** = $\binom{8}{5}$ magnitude = 9.43

(b) 4**c** = $\binom{-8}{8}$ magnitude = 11.31

(c) 2**d** + **b** − **d** = **d** + **b** = $\binom{7}{-3}$ magnitude = 7.62

(d) n**a** + m**a** = $(n + m)$**a** = $\binom{1}{5}$ magnitude = 5.10

(e) a + 3**d** − **c** − 2**a** = 3**d** − **c** − **a** = $\binom{1}{-16}$

magnitude = 16.03

2 (a) $\binom{2}{5}$ **(b)** $\binom{6}{-1}$ **(c)** $\binom{4}{-6}$

(d) $\binom{-4}{6}$ **(e)** $\binom{4}{2}$

Transformations [p 91]

1 (a) (i) (4, 6), (4, 8), (3, 8)
(ii) (9, 3), (9, 5), (8, 5)

(b) (i) (6, 0), (6, 2), (7, 2)
(ii) (2, 3), (2, 4), (0, 4)

(c) (i) (−7, −3), (−5, −3), (−5, −2)
(ii) (0, −4), (0, −6), (1, −6)

(d) (i) (2, −5), (2, −1), (0, −1)
(ii) (0, −10), (0, −4), (−3, −4)

2 (a) Translation with vector $\binom{-3}{0}$

(b) Reflection in $y = -3$

(c) Rotation 90° anticlockwise, centre (−2, 1)

(d) Enlargement, scale factor 4, centre (7, 0)

Handling data

Pie charts [p 93]

1

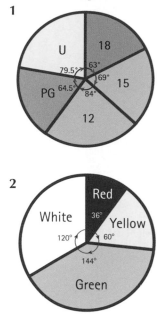

2

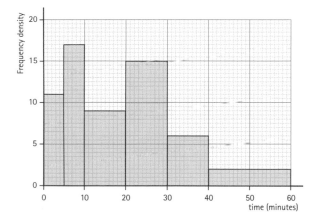

Histograms [p 95]

1 (a)

frequency density / time (minutes)

(b) 20 students were absent.

2 (a) and **(b)**

Height (h cm)	Frequency density	Frequency
$0 \leqslant h < 20$	0.15	3
$20 \leqslant h < 30$	2.2	22
$30 \leqslant h < 40$	2.7	27
$40 \leqslant h < 50$	3.1	31
$50 \leqslant h < 55$	2.8	14

(c) Graham has 97 plants.

Finding averages [p 97]

1 mode: 12

median: 12.5

mean: 13.5

range: 7

2 mode: 25 g

median: 25 g

mean: 25.2 g

range: 4 g

3 mode: $10 \leqslant x < 15$

median: $10 \leqslant x < 15$

mean: £11.92 (estimated)

range: 40 (estimated)

Cumulative frequency [p 99]

1

Age, A years	Cumulative frequency
$A < 10$	8
$A < 20$	17
$A < 30$	28
$A < 40$	35
$A < 50$	50
$A < 60$	72
$A < 80$	86
$A < 100$	88

2

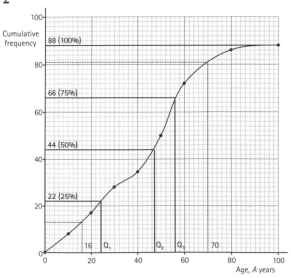

3 median = 47 years

lower quartile = 24 years

upper quartile = 56 years

interquartile range = 32 years

4

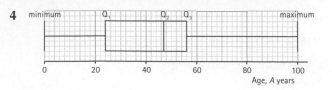

5 (a) $88 - 81 = 7$ people

(b) $\frac{13}{88} = 0.15$ (2 dp)

Comparing sets of data [p 101]

1 (a)

	longest	shortest	mean	range
Markit	1730	1650	1700	80
Skribbla	1930	1640	1750	290

(b) Skribbla: the mean is slightly higher, the shortest is about the same and the longest is quite a bit longer. The poorer consistency turns out to be a good thing!

2 (a) 139 in total

(b) The Gromore results are slightly more consistent and have a higher mode than the Sproutwell plants, so Gromore appears to be better.

Probability [p 103]

1 Total number of cars counted = $144 + 75 + 36 + 33 + 12 = 300$

(a) $P(1) = \frac{144}{300} = \frac{12}{25}$ or 0.48

(b) $P(\geqslant 3) = \frac{36 + 33 + 12}{300} = \frac{81}{300} = \frac{27}{100}$ or 0.27

2 See diagram on opposite page.

3 (a)

		\multicolumn{6}{c}{1st die}

		1	**2**	**3**	**4**	**5**	**6**
2nd die	**1**	0	1	2	3	4	5
	2	1	0	1	2	3	4
	3	2	1	0	1	2	3
	4	3	2	1	0	1	2
	5	4	3	2	1	0	1
	6	5	4	3	2	1	0

(b) 1 is most likely: $P(1) = \frac{10}{36}$

(c) $P(2) = \frac{8}{36} = \frac{2}{9}$. $\frac{2}{9}$ of $100 = 22.22...$, so roughly 22 times.

Answer to 2, Probability

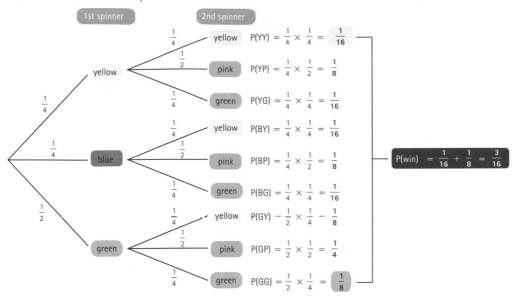

Scatter diagrams and correlation [p 105]

1

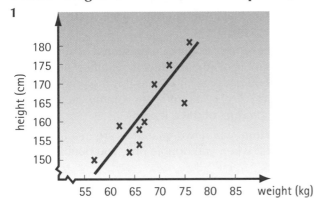

2 There is a strong correlation between height and weight.

3 Nisha's height will be about 152 cm.

Glossary

A

acute angle an angle less than 90° (less than a right angle)

algebra the branch of mathematics that deals with the general case. Algebra involves the use of letters to represent variables and is a very powerful tool in problem-solving.

alternate segment theorem This states that the angle between a chord and the tangent to the circle at one of its endpoints is equal to the angle subtended by the chord in the segment opposite.

angle a measure of rotation

apex the 'point' of a pyramid

arc part of the circumference of a circle

area a measure of two-dimensional space using square units, e.g. cm²

average see mean, median and mode

B

bar chart a frequency chart, where the frequency of the data is proportional to the height of the bar

base the large number in a power – the number being repeatedly multiplied

bearing a measure of direction used in navigation

BIDMAS a made-up 'word' to help remember the order of operations – **B**rackets, **I**ndices, **D**ivision/**M**ultiplication, **A**ddition/**S**ubtraction

box and whisker diagram a graphical representation of the minimum and maximum values in a data set (the whiskers) and the median and quartiles (the box)

C

centre of enlargement/rotation the point that is fixed during enlargements/rotations

chord a line across a circle, not passing through the centre

circumference the complete boundary of a circle

coefficient the number in front of the variable that multiplies it, for example in $2x$ the coefficient of x is 2 and in $5y$ the coefficient of y is 5

common factor a factor of two or more numbers

common multiple a multiple of two or more numbers

composite number any number that is not prime

compound interest the type of interest that is paid by most banks, where the interest is added to the principal invested and then in subsequent years interest is earned by the original interest

compound unit a unit built from more than one basic unit, such as m/s (speed) or kg/m³ (density)

congruent shapes shapes that are identical. Rotations and reflections are ignored when checking congruence

construction precise mathematical drawing using a compass, ruler and protractor

correlation the connection between two variables. Correlation is positive if, as one value increases, the other increases; negative if, as one increases, the other decreases.

cosine in a right-angled triangle, equal to adjacent ÷ hypotenuse

cosine rule an extension of Pythagoras' Rule to all triangles: $c^2 = a^2 + b^2 - 2ab \cos C$.

cube (1) a cuboid whose edges are all of equal length

cube (2) the third power of a number, e.g. 2^3 (two cubed) = 8

cube root the opposite of cubing, e.g. $\sqrt[3]{64} = 4$ because $64 = 4^3$

cuboid a rectangular prism, having six rectangular faces

cumulative frequency Think of this as a running total graph. Add up the frequencies as you go through the data. The cumulative frequency is drawn as a typical 'S' shaped graph. The curve is also known as an ogive.

cylinder a circular prism

D

decomposition writing a number as the product of its prime factors, e.g. $120 = 2^3 \times 3 \times 5$

denominator the lower part of a fraction

diameter the distance across a circle from one point on the circumference to another point on the circumference, passing through the centre

dimension the type of quantity represented by an expression, built up as the product of lengths. For example, xy could represent area as it is the product of two lengths; $xyz + r^3$ could indicate volume, as both terms are the product of three lengths.

direct proportion when two quantities are in a fixed ratio to one another: when you double one, you double the other, etc.

distribution a set of data

E

elevations views of a solid object from the front and side

equation a mathematical statement, usually in algebra, where two sides of the statement are equal. The aim is to work out the value of the unknown by manipulating the equation.

equation, linear an equation where the highest power of any of the terms is 1

equation, quadratic an equation where the highest power of any of the terms is 2

equilateral triangle a triangle that has all three sides of equal length and all three angles equal 60°

estimate the process of comparing the size of a property of an object with a known quantity

expansion removing brackets from algebraic expressions by multiplying out

F

factor a number or variable that divides into other numbers or variables without a remainder

factorisation the process of extracting the highest common factors from an expression

formula an algebraic statement that gives instructions for calculating its subject. For example, $A = \pi r^2$ is accepted as the formula for the area of a circle. It is not necessary to prove this formula every time you need to use it.

frequency density for a group or class of data within a distribution, the frequency divided by the class width and is equal to the height of the bar in a histogram

frequency polygon a way of displaying grouped data where the midpoints of the class intervals are joined by lines

function a rule which applies to one set of quantities and how they relate to another set

G

generalising a process in mathematical thinking, where a general rule, usually expressed in algebra, is determined

gradient the measurement of the steepness of a line. It is the ratio of the vertical to the horizontal distance. Where a line slopes from bottom left to top right, the gradient is positive. Where the line slopes from top left to bottom right, the gradient is negative.

graph a visual representation of a set of data, or a relationship between sets of data

H

highest common factor the highest factor that will divide exactly into two or more numbers

histogram frequency graph resembling a bar chart for grouped data, in which the frequency of each group or class is represented by the area of its bar

hypotenuse the longest side of a right-angled triangle

I

identity an algebraic statement where the two sides are equal regardless of the values you substitute; for example $2x = x + x$

image the result of a transformation acting on an object shape

imperial units the traditional units of measure which were once used in the UK and the USA, such as feet, inches, pints and gallons

improper fraction a fraction where the numerator is of a higher value than the denominator

inclusive inequality an inequality in which the endpoint is included, e.g. $x \leqslant 5$

independent event an event whose outcome is not dependent on the outcomes of other events

index the small number in a power – how many copies of the base to multiply together

inequality an algebraic statement which is unequal

integer a whole number

intercept the position where a graph crosses the y-axis

interquartile range the spread of the middle 50% of a distribution

irrational number a number that cannot be written as a fraction, for example $\sqrt{2}$ or π

isosceles triangle a triangle with two equal sides and two equal angles, and where the equal sides are opposite the equal angles

L

line of symmetry a line that bisects a shape so that each part of the shape reflects the other

locus a set of positions governed by a mathematical rule. Loci may be points, lines, curves or areas.

lowest common multiple the lowest number that two or more numbers will divide into; for example the lowest common multiple of 4 and 8 is 8, because 8 is the lowest number that both 4 and 8 will go into

M

mean the arithmetic average, which is calculated by adding all of the items of data and then by dividing the answer by the number of items of data

median the middle value in a set of data which are in order of size from highest to lowest or lowest to highest

metric the system of measurements usually used in Europe and the most commonly used system in science

mode the value that occurs the most often in a set of data

N

negative number a number with a value less than zero

net the pattern made by a three-dimensional shape when it is cut into its construction template and then laid flat

number, cube the sequence of cube numbers is 1, 8, 27, 64, 125, etc. It is made up from $1 \times 1 \times 1, 2 \times 2 \times 2, 3 \times 3 \times 3$ and so on.

number, prime a number with two and only two factors. The factors are the number itself and 1.

number, square the sequence of square numbers is 1, 4, 9, 16, 25, 36, etc. It is made up from $1 \times 1, 2 \times 2, 3 \times 3$ and so on.

numerator the upper part of a fraction

O

object 'original' shape, before a transformation is applied

obtuse refers to any angle greater than 90° but less than 180°

outcome one possible result of a chance event

P

parallel lines lines that are equidistant along the whole of their length, in other words they are a constant distance apart along the whole of their length

perimeter the total distance around the boundary of a shape

period (1) the number of repeating digits in a recurring decimal

period (2) the number of degrees contained in one repeating section of a trigonometric graph

perpendicular lines lines that meet at right angles

pi (π) the ratio of the circumference of a circle to its diameter, approximately 3.142

pie chart a chart in the shape of a circle, where the size of the sector shows the frequency

plan a view of a solid object from above

polygon a many-sided shape

position-to-term rule a formula that calculates the terms (u_n) of a sequence from their positions (n)

possibility space table a way of representing the outcomes of two events, if these outcomes are equally likely

power the result of multiplying a base number by itself several times, e.g. $5^3 = 5 \times 5 \times 5 = 125$

prefix a short word added to the front of a unit to change its size, e.g. milli-, centi-, kilo-, etc.

prism a solid with a uniform cross-section

probability the study of the chance of events occurring

proportional division dividing an amount in a given ratio

proportionality a relationship of the form $y = kx^n$ for some value of n; also written $y \propto x^n$

proportion, direct a relationship of the form $y = kx$

proportion, inverse a relationship of the form $y = \frac{k}{x}$

Pythagoras' Rule (theorem) the statement of the relationship between the hypotenuse (h) of a right-angled triangle and the other two sides (a and b). Using algebra, it is written $h^2 = a^2 + b^2$

Q

quadrant one of the four 'quarters' of a co-ordinate grid

quadrilateral a plane four-sided shape

R

radius the distance from the centre of a circle to any point on the remaining part of the circumference. $2 \times$ radius = diameter.

ratio a proportional relationship between one quantity and another, e.g. if two amounts are in the ratio 3 : 1, the first is always 3 times the second

rational number a number that can be written as a fraction

reflex angle any angle greater than 180°.

regular polygon a plane shape having straight sides of equal length and equal interior angles

right angle any angle that is equal to 90°

rounding reducing the accuracy of a number to a sensible level. Rounding may be to the nearest integer, ten, hundred, etc., to a specified number of decimal places or significant figures.

S

scale factor the number of times a shape is enlarged

scalene triangle a triangle where all of the sides and all of the angles are of different sizes

scatter diagram a plot of two sets of data values as co-ordinates, used to investigate a possible link between them

sequences a set of numbers having a common property. It is possible to work out what the numbers have in common and so predict further numbers in the sequence.

significant figures or digits the most important digits in a number – those in the places with the highest values, e.g. in 2156 the 2 is the first significant figure as it is in the thousands place

similar shapes shapes that are identical apart from their size, i.e. one is an enlargement of the other

sine in a right-angled triangle, equal to opposite ÷ hypotenuse

sine rule in any triangle, links each angle to the length of its opposite side: $\frac{\sin A}{a} = \frac{\sin B}{b} = \frac{\sin C}{c}$. In solving a problem, you would normally use just two parts of the rule at a time.

standard form a way of writing very large or very small numbers, using powers of 10

strict inequality an inequality in which the endpoint is not included, e.g. $x > 5$

subject the quantity calculated by a formula, usually on the left of the equals sign

substitution replacing letters in an algebraic expression by their values

supplementary adding to make 180°

surd an expression of the form $a + b\sqrt{c}$, where a, b and c are integers (a may be zero)

T

tangent (1) a straight line that touches a curve at one point.

tangent (2) in a right-angle triangle, equal to opposite ÷ adjacent.

term-to-term rule a formula that calculates the next term of a sequence from the previous term

tolerance the size of the possible error in a measurement, e.g. if $L = 25$ cm, to the nearest cm, then 24.5 cm $\leq L < 25.5$ cm. The tolerance is 0.5 cm.

translation a sliding transformation

tree diagram a way of representing the probabilities for multiple events, where the outcomes are not equally likely

trial and improvement finding an approximate solution to an equation by trying a suitable value, then using the result to obtain a better approximation

U

unitary ratio a ratio in the form 1 : n or n : 1

V

VAT Value Added Tax.

vector a quantity describing a translation, e.g. $\begin{pmatrix} 5 \\ -2 \end{pmatrix}$ means '5 units right and 2 down'

volume a measure of three-dimensional space, using cubic units, e.g. cm^3

Last-minute learner

Number

Types of number
- **Integers** are positive and negative whole numbers. Any integer can be written as a product of **prime factors**,
  e.g. $24 = 2^3 \times 3$, $450 = 2 \times 3^2 \times 5^2$.
- **Rational numbers** include integers and fractions. Numbers like $\sqrt{2}$, $\sqrt[3]{10}$ and π are **irrational**.

Indices
- **Index laws:** in algebra, $a^n \times a^m = a^{(n+m)}$, $a^n \div a^m = a^{(n-m)}$.
- A negative power is just the **reciprocal** of the positive power. $2^{-2} = \frac{1}{2^2} = \frac{1}{4}$, $2^{-3} = \frac{1}{2^3} = \frac{1}{8}$, etc.
- **Fractional** indices mean **roots**.
- Numbers in **standard index form** consist of a power of 10 multiplying the significant digits of the number.
  2 million $= 2 \times 1\,000\,000 = 2 \times 10^6$.
  $2\,500\,000 = 2.5$ million $= 2.5 \times 10^6$.
- Numbers less than 1 need a negative index,
  e.g. $0.002\,13 = 2.13 \times 10^{-3}$.

- You may be required to leave some answers in **surd** form. Use facts such as $\sqrt{a} \times \sqrt{b} = \sqrt{ab}$.
- Rationalise denominators of types $\sqrt{a}$ and $(a + \sqrt{b})$ by multiplying top and bottom by $\sqrt{a}$ and $(a - \sqrt{b})$, respectively.

Fractions and decimals
- To convert a fraction to its equivalent decimal, divide numerator by denominator.
- To convert a terminating decimal to a fraction, use a sufficiently large power of 10 as denominator. For recurring decimals, multiply by a power of 10 determined by the period of the decimal, then subtract to eliminate the recurring part.

Proportional quantities
- If y is **directly proportional** to x, this is written $y \propto x$. That means $y = kx$, where k is the **constant of proportionality**. Two proportional amounts plotted against each other on a **graph** give a straight line through the origin. If y is proportional to a power of x, this is written $y \propto x^n$, meaning $y = kx^n$.
- If y is **inversely proportional** to a power of x, this is written $y \propto \frac{1}{x^n}$, meaning $y = \frac{k}{x^n}$.

Algebra

Formulae and expressions
- **Substitution** is replacing letters in a formula, equation or expression by numbers (their **values**). Be careful to evaluate parts of the formula in the correct order.
- **Expressions** in algebra are made up of a number of terms added or subtracted together. Each **term** is made up of letters and numbers multiplied or divided together. Combine **like terms** to simplify an expression.
- A **formula** usually has its **subject** on the left-hand side of the equals sign and an expression on the right-hand side. Any letter in a formula can become the subject by **rearranging** it. As long as you do the same thing to both sides of your formula, it is still true.

Multiplying and dividing terms
- When **multiplying two terms** together, multiply the numbers first, then multiply the letters in turn, using the index rules.
- When **dividing terms**, write the question in fraction style if it's not already written that way and cancel the numbers as if you were cancelling a fraction to lowest terms, then divide the letters in turn.

Expanding brackets
- When a number or letter multiplies a bracket, **everything** inside the bracket is multiplied. Removing the brackets is called **expanding** them.

Factorisation
- **Factorising** is the opposite of expansion. To **factorise**, look for **common factors** between the terms.

This process is called **extracting factors**. Sometimes you need to do this in more than one step.
- **Quadratic** expressions can sometimes be factorised into two brackets. First, write down a list of the numbers that could be part of the x^2 term. Write down a list of the numbers that could be part of the number term. Test combinations of these numbers to see if you can match the x term in the expression you want to factorise.
- **Difference of two squares**:
  $a^2 - b^2 = (a - b)(a + b)$.

Quadratic equations
- Quadratic equations of the form $ax^2 + bx + c = 0$ can be solved by factorisation, or by using the quadratic formula:
  $$x = \frac{-b \pm \sqrt{b^2 - 4ac}}{2a}.$$

Trial and improvement
- Sometimes you can't find an exact solution to an equation but can find a reasonable **approximation** using **trial and improvement** or a **decimal search**. You use the results of a 'guess' to make better guesses.

Simultaneous equations
- **Simultaneous equations** are pairs of equations with two unknown letters that are both true at the same time. The technique of **elimination** involves adding or subtracting the equations so that one of the letters disappears (is **eliminated**). Sometimes, you have to multiply one or both of the equations. This is to match coefficients in order to add or subtract and eliminate.

- Another method of solution is to express one variable in terms of the other, then **substitute** the resulting expression into the other equation. This is useful when one equation is quadratic.
- Each equation in a simultaneous pair has its own **graph**. The x and y co-ordinates of the point where the graphs **intersect** give the solution.

Inequalities
- Ranges of numbers are described using **inequalities**.
- There are four inequality symbols:

$>$ greater than	$\geqslant$ greater than or equal to
$<$ less than	$\leqslant$ less than or equal to

- 'All the numbers that are 3 or less' is described by the inequality $x \leqslant 3$.
- Sometimes inequalities can be combined. Suppose that $x < 2$ and $x \geqslant -3$. This makes a **range inequality**, $-3 \leqslant x < 2$.
- Solve inequalities using the techniques for equations, e.g. the solution of $2x - 1 > 5$ is $x > 3$.
- A line (e.g. $y = 5 - x$) divides a co-ordinate grid into two **regions**. The region above the line is $y > 5 - x$ and the one below is $y < 5 - x$. The line can be included in the region by using $\geqslant$ or $\leqslant$.

Sequences and functions
- **Sequences** are made up of a succession of **terms**. Each term has a **position** in the sequence: 1st, 2nd, etc. A **linear** sequence has the formula $u_n = an + b$.
- **Quadratic** sequences have formulae of the form $u_n = an^2 + bn + c$. To analyse quadratic sequences, look at the **second** difference row.
- A **function** is a rule applied to a variable, e.g. if $f(x) = x^2 - 3$, then $f(5) = 22$ and $f(x + 1) = (x + 1)^2 - 3 = x^2 + 2x - 2$.

Transformations of graphs
The graph of $y = f(x)$ is transformed as follows:
$y = f(x) + a$: translation a units in the y direction.
$y = f(x + a)$: translation $-a$ units in the x direction.
$y = af(x)$: stretch in the y direction, scale factor a.
$y = f(ax)$: stretch in the x direction, scale factor $\frac{1}{a}$.

Shape, space and measures

Accuracy of measurements
- Suppose the length L of a piece of wire is 67 mm, to the nearest mm: then $66.5 \leqslant L < 67.5$.
  If L is given to the nearest 0.1 mm (1 dp) instead, then $66.95 \leqslant L < 67.05$.

Speed, distance and time
- Units of speed are metres per second (m/s), kilometres per hour (km/h), miles per hour (mph), etc.
- Use the 'd-s-t triangle'. Cover up the letter you want to work out: the triangle gives the formula.
- On a distance-time graph, the gradient of the tangent gives the speed at a given time.
- On a velocity-time graph, the gradient of the tangent gives the acceleration at a given time, and the area bounded by the graph and the time axis gives the distance travelled.

Mensuration – length and distance
- In any right-angled triangle, the **hypotenuse** is the side opposite the right angle.
- Pythagoras' Rule: in any right-angled triangle, $h^2 = a^2 + b^2$. Use it to calculate the diagonal of a rectangle, or the distance between two points on a co-ordinate grid.

- The **space diagonal** of a cuboid with dimensions a, b and c is $d = \sqrt{a^2 + b^2 + c^2}$.
- The **circumference** of a circle of diameter d (radius r) is $C = \pi d = 2\pi r$. $\pi \approx 3.142$.
  The length of an arc subtending $x°$ at the centre is $\frac{x}{360}$ of the circumference.
- When a wheel turns once (makes one **revolution**), the distance moved by whatever it's attached to (e.g. a car, bike, etc.) is the same as the circumference of the wheel.

Trigonometry
- In a right-angled triangle, $\sin x = \frac{\text{opposite side}}{\text{hypotenuse}}$.
  $\cos x = \frac{\text{adjacent side}}{\text{hypotenuse}}$, $\tan x = \frac{\text{opposite side}}{\text{adjacent side}}$
- To find an angle in a right-angled triangle with known sides, calculate the trig ratio, then use the **inverse trig function** ($\sin^{-1}$, etc.).
- In any triangle ABC with sides a, b and c opposite their corresponding vertices, you can use:
  - the sine rule: $\frac{a}{\sin A} = \frac{b}{\sin B} = \frac{c}{\sin C}$;
  - the cosine rule: $a^2 = b^2 + c^2 - 2bc \cos A$ (also $b^2 = c^2 + a^2 - 2ca \cos B$ and $c^2 = a^2 + b^2 - 2ab \cos C$).

Area

- Using l for length and w for width, the area A of a rectangle is given by $A = lw$.
- Triangles, parallelograms and trapezia all share an important measurement: the **perpendicular height**. These are the area formulae:

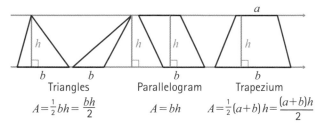

Triangles
$A = \frac{1}{2}bh = \frac{bh}{2}$

Parallelogram
$A = bh$

Trapezium
$A = \frac{1}{2}(a+b)h = \frac{(a+b)h}{2}$

- The area of the circle of radius r is $A = \pi r^2$. The area of a sector subtending $x°$ at the centre is $\frac{x}{360}$ of the total area.
- In any triangle ABC as described above, the area is given by $A = \frac{1}{2}ab \sin C$.

Surface area

- The surface area of a solid object is the combined area of all the faces on the outside. Curved surfaces on spheres, cones and cylinders form part of the surface area too.
- For a prism with cross-section of perimeter P and area A, the total surface area $S = 2A + Pl$. For a cylinder, $S = 2\pi r^2 + 2\pi rl = 2\pi r(r + l)$.
- Cones need an extra measurement, the slant height s. The curved surface is πrs, so the total surface area is $\pi r^2 + \pi rs = \pi r(r + s)$.
- The surface area of a sphere is $4\pi r^2$.

Volume

- There are four basic volume formulae.

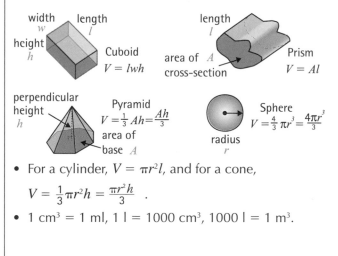

Cuboid
$V = lwh$

Prism
area of cross-section A
$V = Al$

Pyramid
$V = \frac{1}{3}Ah = \frac{Ah}{3}$
area of base A

Sphere
$V = \frac{4}{3}\pi r^3 = \frac{4\pi r^3}{3}$
radius r

- For a cylinder, $V = \pi r^2 l$, and for a cone,

$V = \frac{1}{3}\pi r^2 h = \frac{\pi r^2 h}{3}$.

- $1 \text{ cm}^3 = 1 \text{ ml}$, $1 \text{ l} = 1000 \text{ cm}^3$, $1000 \text{ l} = 1 \text{ m}^3$.

Angles and Shapes

- Whenever lines meet or **intersect**, the angles they make follow certain rules.

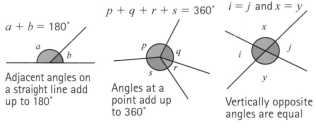

Adjacent angles on a straight line add up to 180°
$a + b = 180°$

Angles at a point add up to 360°
$p + q + r + s = 360°$

Vertically opposite angles are equal
$i = j$ and $x = y$

- Three types of relationship between angles are produced when a line called a **transversal** crosses a pair of **parallel** lines.

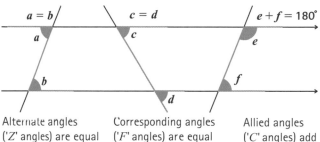

Alternate angles ('Z' angles) are equal
$a = b$

Corresponding angles ('F' angles) are equal
$c = d$

Allied angles ('C' angles) add up to 180°
$e + f = 180°$

- The exterior **angles** of a polygon always add up to exactly 360°.
- Every type of polygon has its own **interior angle sum**. You can calculate it using any of these formulae:
  n is the number of sides and S is the angle sum.
  $S = (n - 2) \times 180°$ $S = (180n - 360)°$
  $S = (2n - 4)$ right angles
- Work out the interior angles for **regular** polygons in two ways: work out the angle sum, then divide by the number of sides; or divide 360° by the number of sides to find one exterior angle, then take this away from 180°.

Transformations

- Mathematical transformations start with an original point or shape (the object) and transform it (into the **image**).
- A **translation** is a 'sliding' movement, described by a **column vector**, e.g. $\begin{pmatrix} 5 \\ -4 \end{pmatrix}$.
- In a rotation, specify an angle and **centre of rotation**.
Given a rotation, to find the centre:
- join two pairs of corresponding points on the object and image
- draw the perpendicular bisectors of these lines
- the point of intersection is the centre of rotation.
- Reflection in any line is possible, but the most likely ones you will be asked to use are these:
  - horizontal lines ($y = a$ for some value of a);
  - vertical lines ($x = b$ for some value of b);
  - lines parallel to $y = x$ ($y = x + a$ for some value of a);
  - lines parallel to $y = -x$ ($y = a - x$ for some value of a).

- To describe an enlargement you need to give a **scale factor** and a centre of enlargement. To find the **centre of enlargement**, draw lines through corresponding points on the object and image. These all intersect at the centre of enlargement. Enlargements are mathematically **similar** to their objects.

Congruence tests

If the following features of two triangles match, the triangles are congruent: 3 sides (**SSS**); 2 sides and the included angle (**SAS**); 2 angles and a side (**AAS**); in a right-angled triangle, the hypotenuse and one other side (**RHS**).

Loci

A set of positions generated by a rule is called a **locus**. The four major types are as follows:

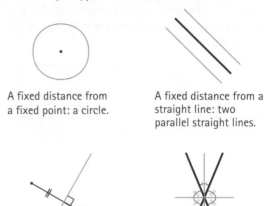

A fixed distance from a fixed point: a circle.

A fixed distance from a straight line: two parallel straight lines.

Equidistant from two fixed points: the perpendicular bisector of the points.

Equidistant from two straight lines: the bisectors of the angles between the lines.

Often, you need to combine information from two or more loci. This will lead to a region or area, a line segment, or one or more points.

Handling data

Histograms

- In a **histogram**, **frequency density** (frequency ÷ class width) is plotted against data value. This has the effect that equal frequencies in the data are represented by equal areas on the histogram.

Averages

- The **mode** is the most common value. With grouped data, the group with the highest frequency is called the **modal** group or class. The **median** is the middle value in a set, when all the numbers are arranged in order. The **mean** is the sum of the data items, divided by the total frequency.

Spread

- The **range** is simply the difference between the smallest and largest data items. The **interquartile range** encloses the middle 50% of a data set.

Probability

- **The OR rule**: when two outcomes A and B of the same event are exclusive, $P(A \text{ or } B) = P(A) + P(B)$.
- **Theoretical probability** is calculated by analysing a situation mathematically.
- **Experimental probability** is determined by analysing the results of a number of trials of the event.
- **The AND rule**: when two events X and Y are independent, $P(X \text{ and } Y) = P(X) \times P(Y)$.
- Use a **tree diagram** to organise multiple events with varying probabilities.

Notes

Notes